8 Progress in Clinical Biochemistry and Medicine

Clinical Biochemistry in Hepatobiliary Diseases

Editors:
F. Salvatore, A. Roda, L. Sacchetti

Proceedings of the International Satellite Symposium, Bologna, Italy, 1988

With Contributions by
C. Armanino, Y. Artur, N. Blanckaert, G. Castaldo, D. Festi, J. Fevery, M. M. Galteau, J. Griffiths, A. Minutello, D. W. Moss, M. Muraca, I. W. Percy-Robb, R. Rizzoli, A. Roda, E. Roda, S. B. Rosalki, L. Sacchetti, F. Salvatore, G. J. Sanderink, F. Schiele, G. Siest, P. Simoni, F. Vanstapel, M. Wellman, M. Werner

With 72 Figures

Springer-Verlag Berlin Heidelberg New York
London Paris Tokyo Hong Kong

Guest Editors

Prof. Francesco Salvatore

Dipartimento di Biochimica e Biotecnologie Mediche, Università di Napoli, II Facoltà di Medicina, Via S. Pansini, 5, I-80131 Napoli/ Italy

Prof. Aldo Roda

Istituto di Chimica Analitica, Università di Messina/Italy

Prof. Lucia Sacchetti

Dipartimento di Biochimica e Biotecnologie Mediche, Università di Napoli, II Facoltà di Medicina, Via S. Pansini, 5, I-80131 Napoli/ Italy

ISBN-13: 978-3-642-74396-2 e-ISBN-13: 978-3-642-74394-8
DOI: 10.1007/978-3-642-74394-8

Softcover reprint of the hardcover 1st edition 1989

Bookbinding: Lüderitz & Bauer, Berlin
2151/3020-543210 — Printed on acid-free paper

Editorial Board

Preface

The clinical biochemistry of hepatobiliary diseases is very widely studied, and publications abound on this topic. However, there is no recent publication that provides a comprehensive collection of the various leading aspects that go to make up this complex theme. Therefore, we thought it useful to gather together a few scientists whose work has focused on the various clinical biochemistry aspects of these disorders in order that they might discuss their experience and expertise.

The aim of the International Satellite Symposium on Clinical Biochemistry in Hepatobiliary Disease, in addition to reviewing the individual aspects, was to describe the state-of-the-art so as to provide useful data for laboratory scientists and also for physicians working in the field of hepatobiliary diseases, and these two aims are clearly reflected in the chapters of this volume.

The volume opens with an introductory chapter that gives a general overview of the various aspects of the clinical biochemistry of these disorders, while the closing chapter deals with an important aspect that deserves to be increasingly emphasized in laboratory medicine, i.e., strategies to integrate information coming from the laboratory to make them more useful for clinical diagnosis.

Despite the widespread use of biochemical tests in hepatobiliary diseases, they are still unable to provide a key with which to decipher the biochemical characteristics of these disorders. The five chapters in this book devoted to clinical enzymology go some way to giving an insight that may provide such a key. One of these chapters concerns a biochemical signal, the serum gamma-glutamyltransferase pattern, which is slowly emerging from the clinical enzymology of hepatic diseases and that might contribute to discriminating between various hepatobiliary disorders. Two chapters are devoted to alkaline phosphatase, one to amylase and another to specific enzymatic systems that are particularly involved in hepatic diseases. This has been an extremely fruitful area over the past few years, and these four chapters admirably review the progress made and provide indications regarding the directions in which this field will develop in the future.

One of the remaining two papers deals with particular aspects of the clinical biochemistry of hepatobiliary diseases that are involved in the metabolism of biliary pigments, particularly bilirubins. This topic has received much attention in the past; however, the advancements made in recent years have transformed the field, and Dr. Blanckaert and coworkers have provided a lucid update of the present situation. Lastly, there is the chapter devoted to cholesterol catabolism, particularly as regards bile acids. Here Professor Roda and his group give an extensive overview of both the chemical-physical and clinical aspects of bile acid analysis in biological fluids.

We would like to end this brief preface by thanking Jean Gilder for having revised and edited the papers where necessary and for having compiled the subject index. In addition, we are most indebted to Boehringer Biochemia Robin (Italy) for their generous support of this symposium.

Italy, February 1989

Francesco Salvatore, MD, Ph.D
Aldo Roda, Ph.D.
Lucia Sacchetti, Ph.D.

Table of Contents

The Clinical Biochemistry of Hepatobiliary Diseases

I. W. Percy-Robb

Professor in Pathological Biochemistry, Department of Pathological Biochemistry, Western Infirmary, Glasgow G11 6NT, UK

Biochemical tests of liver function are a part of the routine investigation of patients who give a history which suggests that liver disease is present.

The transport function of the liver is usually assessed by measurements of total bilirubin concentration in the patient's plasma. The enterohepatic circulation of bile acids provides an endogenous background against which transport can be assessed, the molar ratio of bile acids to bilirubin traversing the liver being about 150:1.

The integrity of hepatic parenchymal cells is most commonly assessed by measurements of the activities of the aminotransferase enzymes in serum. Using radioimmunoassay, mass measurements of the glutathione transferase are available and may show rises in liver disease. These enzymes represent about 5% of the total cytosolic protein in hepatic parenchymal cells and it appears that these immunoassay measurements can provide an extremely sensitive liver function test.

Liver biopsy is both invasive and not without risk to the patient. Measurements of procollagen III peptide may provide a non-invasive method for assessing hepatic fibrosis in alcoholic liver disease and cirrhosis.

The use of bile acid measurements and of high performance liquid chromatography of bilirubin in assessing hepatic transport, of mass measurements of the glutathione transferase, and of procollagen III peptide is reviewed.

Progress in Clinical Biochemistry and Medicine, Vol. 8

1 Introduction

The clinical investigation of patients suspected of having liver disease, or in whom liver disease has previously been established, is based on a complex matrix of data which include not only a carefully taken case history and full clinical examination but in most cases also depends on laboratory tests as well as ultrasound and radiological investigations. The principal aims of this approach are 1) to detect whether liver disease is present, 2) to make a differential diagnosis of the type of liver disease present, 3) to assess the severity of the disease, 4) to follow the progress of the disease and where possible, 5) to assess the likely prognosis. In addition to these laboratory and radiological and ultrasound investigations, the diagnosis in a proportion of cases rests on histological examination of tissue taken at biopsy.

Biochemical tests are used to assess the structural integrity of the liver, its ability to transport substances from the blood into bile and its ability to synthesise and secrete substances into the blood. Many tests of these functions have been advocated, but in clinical practice only a few have proved both informative and feasible to perform on a large scale. The most widely used combination includes the serum total bilirubin concentration as an index of hepatic transport function and of the severity of clinical jaundice, the serum transaminase activity as a measure of the integrity of hepatic parenchymal cells, the serum alkaline phosphatase activity as an index cholestasis and the serum albumin concentration as a measure of hepatic synthetic capacity. Collectively this group of tests are known as "the liver function tests" even though they do not all measure hepatic function. They should, however, not be considered to be routine but rather should be used only when there is some reason to suspect hepatic disorder.

In addition to these widely used tests, the prothrombin time is sometimes used as a test of hepatic synthetic function because it depends on coagulation factors that are synthesised in the liver and the serum globulin concentration may be helpful in assessing the severity of chronic liver disease. A specific set of biochemical tests may be used to detect specific diseases. These include serum alpha-fetoprotein, alpha-1-antitrypsin and ceruloplasmin.

Collectively these biochemical investigations are now widely used and it is not the purpose of this chapter to review their use in detail. Rather, I wish to consider some aspects of three types of investigation which may prove to be useful additional biochemical investigations in patients with liver disease. These are 1) the use of fractionated bilirubin measurements and bile acid measurements especially in the assessment of patients with mild, anicteric hyperbilirubinaemia, 2) the use of immunoassay measurements of the hepatic glutathione-S-transferase enzymes in plasma and 3) the measurement of the procollagen-III-peptide in plasma which may provide a non-invasive method for assessing hepatic fibrosis in alcoholic liver disease and in hepatic cirrhosis.

2 Bilirubin and Bile Acid Measurements

2.1 *Background Physiological Considerations*

One of the commonly recognised clinical features of liver disease is the presence of jaundice. The yellow tinge to sclera and skin leads naturally to clinical curiosity about the plasma bilirubin concentration which has become one of the main components of the group of investigations which together constitute the liver function tests. Bilirubin is transported in the plasma mainly in a reversibly bound form with albumin, its uptake by the liver apparently being carrier mediated. Conjugation of bilirubin at the surface of the endoplasmic reticulum is enzymatic and is followed by excretion of mainly mono- and di-glucuronides of bilirubin in bile. Once secretion of bilirubin conjugates by the biliary tree is completed there is no significant reabsorption of bilirubin from the gut. Bilirubin concentrations in plasma therefore represent the algebraic sum of the rates of entry of bilirubin into the plasma space and of its exit by way of the hepatic uptake process.

Bile acids, on the other hand, are subjected to a rather dynamic enterohepatic circulation, which depends on bile acid secretion by the liver, followed by uptake from

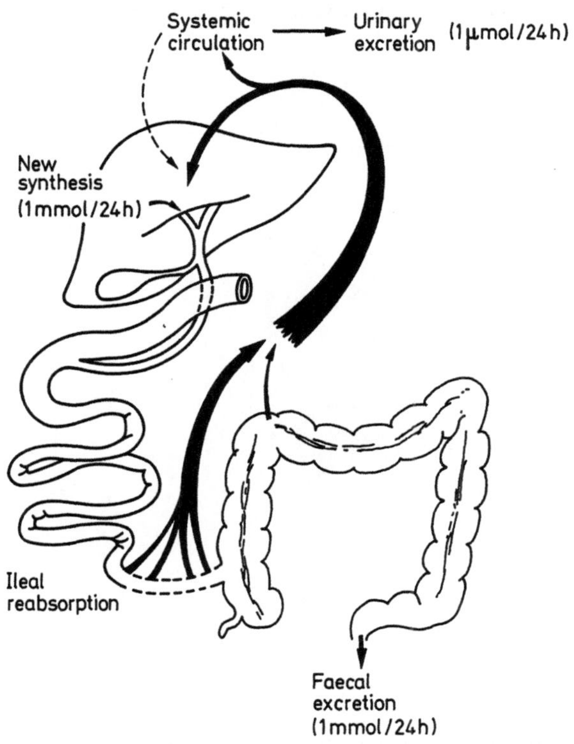

Fig. 1. Diagramatic representation of the enterohepatic circulation of the bile acids in man

the intestinal lumen mainly at the terminal ileum (for a review of the enterohepatic circulation see [1]). In turn these bile acids are delivered into the plasma only to be taken up again by the liver, thereby completing the enterohepatic circulation (Fig. 1) an important feature of which is that it circulates 6–10 times daily thereby contributing a multiplier effect to defects in its essential functional elements both in the terminal ileum and the liver. The effect of this multiplier is best illustrated by comparing the amount of bile acids and of bilirubin that transverse the liver daily. While only about 3 g of bile acids is present in the enterohepatic circulation at any one time, the effect of repeated circulation is that as much as 30 g of bile acids pass across the terminal ileum uptake site and the liver daily. On the other hand, only 200 mg of bilirubin cross the liver each day. Given that bile acids and bilirubin have similar molecular masses this represents a molar excess of bile acids to bilirubin of about 150:1. The enterohepatic circulation of bile acids has been the subject of careful computer modelling [2,3].

The essential driving forces of the enterohepatic circulation in the terminal ileum and liver are not 100% efficient and for this reason bile acids are delivered into the colon with resulting bile acid excretion in faeces (only limited reabsorption from colon occurs), and into the systemic circulation due to incomplete uptake from portal blood by the liver [4,5]. Bile acid concentrations in the systemic circulation show a marked diurnal variation that is closely related to the ingestion of food (Fig. 2). The presence of food in the stomach stimulates gallbladder contraction with a consequent increase in the mass of bile acids that are actively involved in the enterohepatic circulation, rather than being sequestered in the gallbladder.

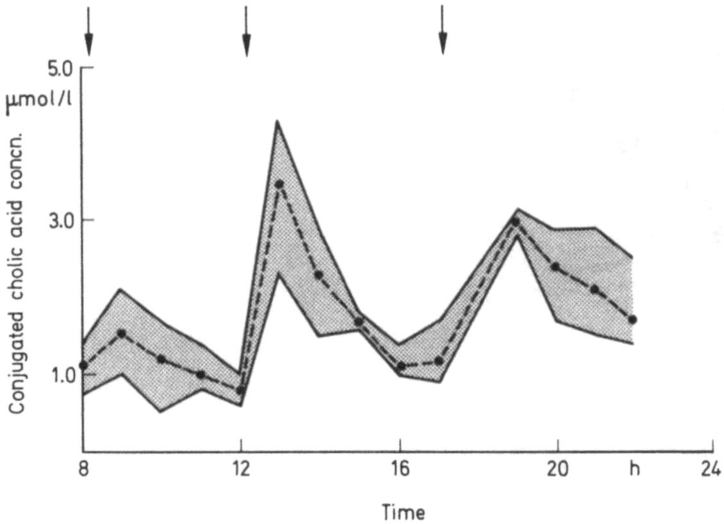

Fig. 2. Fasting and postprandial conjugated cholate concentrations in the serum of seven normal individuals. The arrows indicate the timing of ingestion of normal meals.

2.2 Methods of Bile Acid Measurement in Plasma

Bile acid concentrations can be measured in blood using a variety of different methods which include 1) the use of steroid dehydrogenase enzymes which oxidise the bile acid hydroxyl groups [6], 2) radioimmunoassay [7,8,9] and 3) methods based on gas liquid chromatography [10] (GLC) and high performance liquid chromatography (HPLC). Analytical insensitivity appears to limit the use of many of the enzyme-based assay methods. Radioimmunoassay affords adequate sensitivity to establish the lower limits of the reference interval in normal individuals while fasting (at which time the bile acid concentrations are at their lowest values) but each assay gives data that refer only to one of the different types of bile acids present in plasma. HPLC on the other hand can provide data across the range of bile acids present but is relatively time-consuming to perform and may be limited by analytical sensitivity considerations [11].

2.3 Bile Acid Measurements in Chronic Liver Disease and Terminal Ileal Disease

There is now a large literature reporting the results of bile acid measurements in established liver disease (for review see [12]). It is clear that substantial changes occur in the dynamics of the enterohepatic circulation in liver disease such that bile acid concentrations are increased in the peripheral blood and fall in the lumen of the small intestine to an extent which can adversely effect the absorption of dietary lipid [13].

These findings are consistent with the hypothesis that the bile acid concentration in serum reflects a balance between input, particularly by the active small intestinal uptake mechanism, and output which results from hepatic clearance. Intestinal input is abnormal in patients with bile acid malabsorption such as in ileal resection or ileal disease [14], and this results in reduced bile acid concentrations in serum taken in the postprandial state. On the other hand, the *increased* concentration in serum in liver disease results from a failure of bile acid uptake by the liver either because this function is abnormal in individual hepatocytes or because portal blood is bypassing the liver in chronic liver disease and is therefore not subjected to the hepatic uptake process. In addition, liver disease is one of the risk factors in the production of cholesterol-containing gallstones and in this case the background to these findings probably has its basis, at least in part, in abnormal bile acid secretion by the liver.

However, these abnormalities of bile acid metabolism in chronic liver disease appear not to have an important role in the routine investigation of liver disease and therefore these important aspects of the enterohepatic circulation are not considered to be in the scope of the present article. Rather, the place of bile acid measurements in the clinical biochemistry of patients with mild hyperbilirubinaemia will be considered.

2.4 Bile Acid Measurements in Mild Hyperbilirubinaemia

Mild hyperbilirubinaemia, without abnormalities of the other liver function tests, is the main biochemical finding in the constitutional hyperbilirubinaemia which is

Gilbert's syndrome. A complicating clinical feature of Gilbert's syndrome, however, is the association in some patients with abdominal pain. This, and the recurrent nature of the clinical jaundice, may lead to repeated investigation of these individuals in order to establish whether they have occult liver disease.

Abnormalities of bile salt metabolism can be detected in patients with biopsy-proven liver disease who are anicteric and who have little biochemical evidence of hepatic dysfunction as judged by the more conventional liver function tests [15]. Thus, for example, from 20 anicteric patients who presented consecutively with features suggesting liver disease, a total of nine had raised plasma bilirubin concentrations. In eight of these nine individuals with raised total bilirubin concentrations, the bile acid concentrations too were abnormal. The single exception was a patient with haemosiderosis accompanied by marked hepatic fibrosis in whom not only the fasting and postprandial bile acid concentrations were normal but also the clearance of bile acids given intravenously was normal. Bile acid measurements were abnormal in three patients with normal bilirubin values: in eight patients neither bilirubin nor

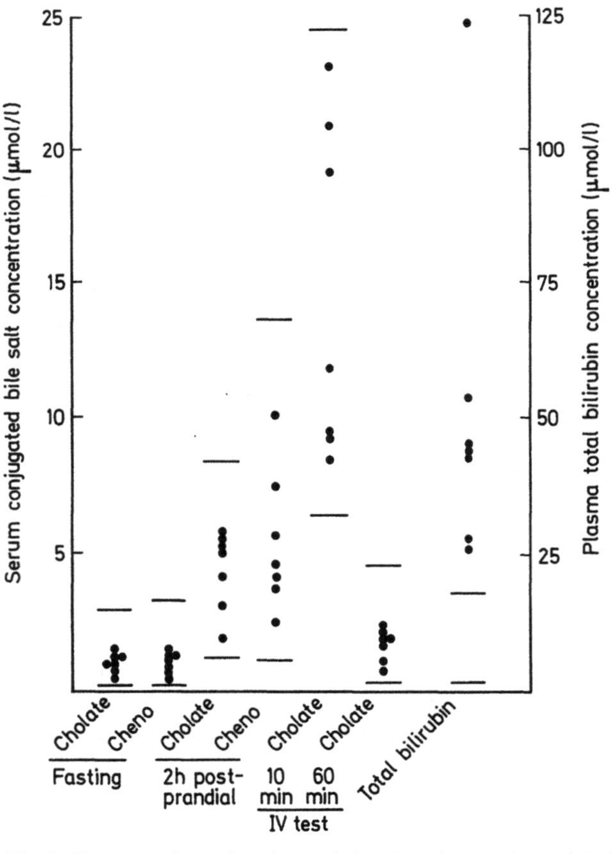

Fig. 3. Concentrations of conjugated cholate and chenodeoxycholate in serum were measured in seven patients with Gilbert's syndrome. Blood was collected in the fasting state, 2 h after ingesting a fat-containing meal and at intervals of 10 and 60 min following an intravenous injection of a bolus of glycocholate (5 μmol/kg body weight).

bile acid concentrations were abnormal. It was concluded from these data that slightly abnormal bilirubin concentrations in patients with liver disease are usually accompanied by abnormal bile acid concentrations.

In Gilbert's syndrome, however, the mild unconjugated hyperbilirubinaemia that is characteristic of the condition is associated not with raised bile acid concentrations but rather with concentrations that are at the *lower* end of the reference interval both in the fasting and the postprandial state (Fig. 3) as well as after intravenous infusion of a bile acid bolus [16,17]. Alternative methods of assessing these patients by measuring the response in plasma bilirubin concentration following caloric restriction [18] or infusion of nicotinamide have been described. These, however, are time-consuming and troublesome to the patient and a single measurement of bile acid concentration either in the fasting or the postprandial state is probably more useful.

2.5 HPLC Assessment of Conjugated/Unconjugated Bilirubin

Chemical methods for the measurement of conjugated and unconjugated bilirubin have long been in use in clinical laboratories. They are, however, intrinsically unreliable. High performance liquid chromatography (HPLC) [19,20] can offer a poten-

4a

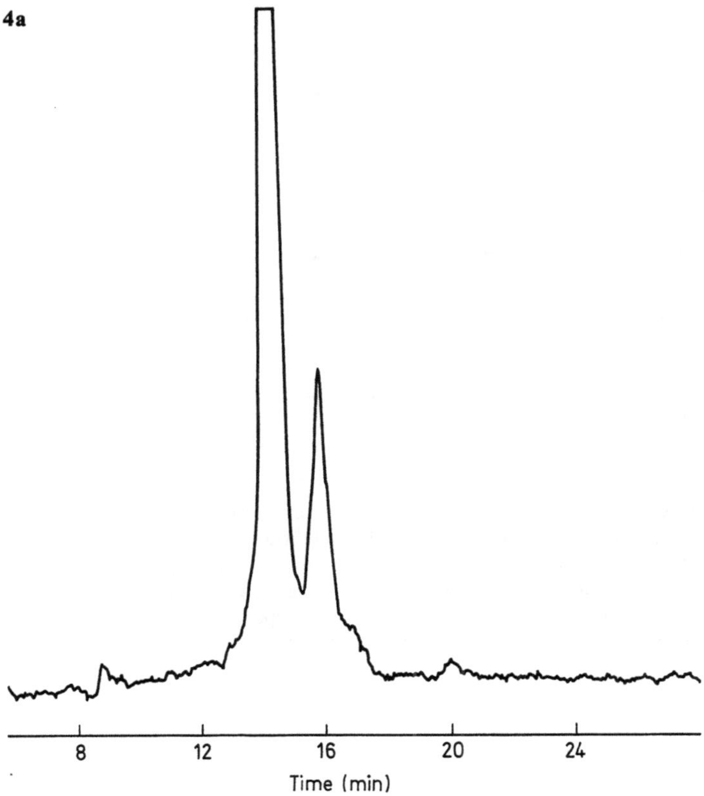

4b

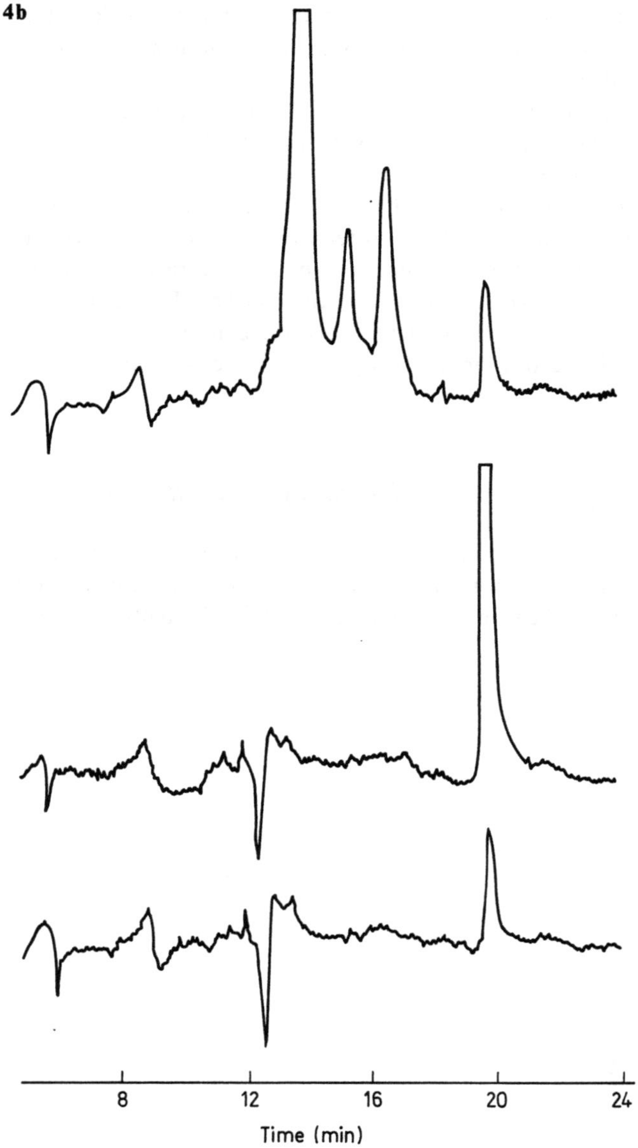

Fig. 4a, b. Bilirubin conjugates in human hepatic bile and in serum were separated using high performance liquid chromatography. Fig.**4a** shows that both bilirubin diglucuronide (retention time 14 min) and monoglucuronide (retention time 16 min) are present. No unconjugated bilirubin is seen. Fig. **4b** shows (lower panel) unconjugated bilirubin in normal serum, (middle panel) raised concentration of unconjugated bilirubin in Gilbert's syndrome and (upper panel) from the left, delta, diconjugated. monoconjugated and unconjugated bilirubin respectively tial solution to this problem by providing bilirubin analyses with effective separation of the di- and mono-glucuronides of bilirubin, the unconjugated form and a fourth form which appears to be bilirubin covalently bound to plasma protein [21]. Examples of the separation achieved using HPLC in human bile and in plasma from 2 patients with Gilbert's syndrome are shown in Fig. 4. It is evident that the sensitivity of the method is adequate for the analysis of plasma from patients with mild hyperbilirubinaemia. While the value of this form of bilirubin analysis in the investigation of these patients needs to be more fully assessed, it seems likely that these measurements will replace chemical methods that have as their basis the differential solubility of the conjugated and unconjugated forms of bilirubin.

3 Clinical Biochemistry and the Glutathione-S-transferase Enzymes

The glutathione-S-transferase (GST) enzymes are a group of multifunctional proteins which play an important role in the detoxification of electrophilic substances and in the hepatic removal of potentially harmful hydrophobic compounds from the blood [22]. In order to perform these functions the GSTs have well-established enzymatic activity catalysing the conjugation of reduced glutathione with a wide spectrum of electrophiles: the formation of glutathione conjugates is the first step in the pathway leading to the formation of mercapturic acids. Binding of hydrophobic compounds such as bilirubin and bile acids to the GSTs has been demonstrated [23]. None of these ligands is known to act as a substrate for the GST activity and binding is therefore taken to play a role in the intracellular processes which result in the hepatic excretion of bile acids and bilirubin in bile. Indeed the protein previously called hepatic ligandin is now known to belong to the GST enzyme group.

The GSTs have a predominantly cystolic distribution in hepatocytes and given that this group of enzymes comprises as much as 3% of the cytosolic protein it is not surprising that the potential use of GST measurements in clinical practice has received attention. While the initial studies were performed using measurements of enzymatic activity in plasma or serum, more recently sensitive radioimmunoassays of the GSTs have been described and it is these that are currently under evaluation [24, 25].

In human liver three distinct groups of GSTs exist. These are the basic, neutral and acidic forms, in each case the enzymes being dimeric proteins with M_r in the range 42,000–46,000 Daltons. The basic group is further divided into at least five forms (αE) and recently it has been shown that two distinct monomers (B_1 and B_2) are present in this group. The neutral and acidic groups too are heterogeneous being separable into at least three subgroups (N/A 1 and N/A 2a and 2b) on the basis of ion-exchange chromatography. A majority of the reports on the clinical use of these assays are based on radioimmunoassays of the B_1B_1 and B_2B_2 homodimers and the neutral acidic fraction N/A 2b. The prime issue at present is one of the sensitivity of the immunoassay in the detection of different forms of liver disease.

3.1 GST in Paracetamol Overdose

Paracetamol-induced hepatotoxicity is a well-established consequence of paracetamol poisoning which can be prevented by administration of compounds such as N-acetyl cysteine (NAC) or methionine. Following the overdose the liver damage becomes rapidly evident clinically, the activities of the aspartate and alanine aminotransferases (AST and ALT) rising rapidly to reach maximum values usually after two to four days. In milder cases of poisoning too the AST and ALT values will usually rise, but the correlation between these changes in enzyme activities in plasma and the liver histology appear to be poor. The decision to use NAC in patients overdosed with paracetamol depends both on the concentration of the drug in the plasma and on the time interval between making the drug measurement and the known time at which the overdose was taken.

Both ALT activities and GST values in plasma rise in paracetamol-induced liver injury [26]. These changes in GST concentrations are present not only in patients in whom clinically recognisable liver injury is evident (both with and without intervention with NAC) but there are small rises in GST in individuals who did not require NAC treatment and in whom the ALT activities do not rise. Examples of the ALT and GST changes in overdosed patients successfully treated with NAC are shown in Fig. 5. The high sensitivity of the changes in GST are clearly shown. In addition,

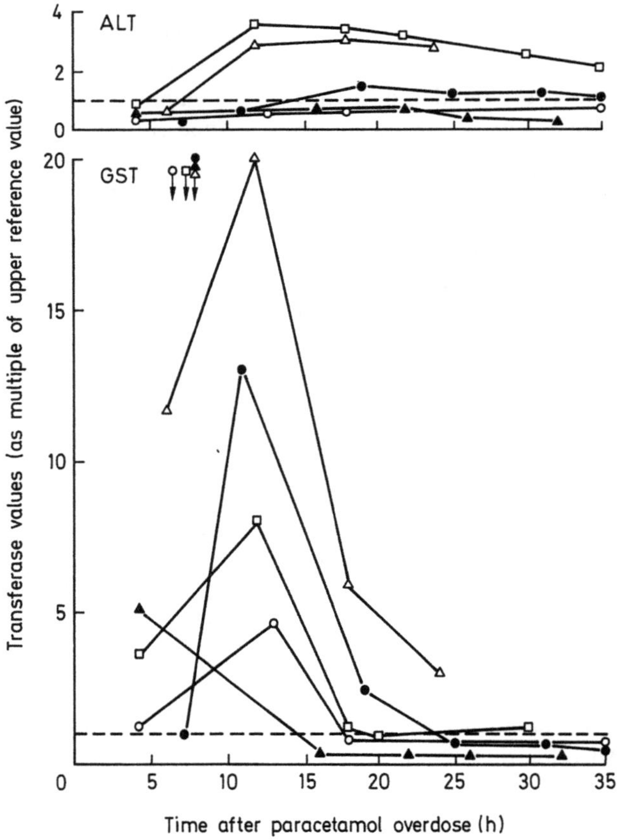

Fig. 5. The activity of alanine amino-transferase (ALT) and the concentration of glutathione-S-transferase (GST) were measured in plasma following overdoses of paracetamol requiring treatment with N-acetyl cysteine which was given at the times given by the arrows

the time course of the raised ALT activities appeared to be longer than those for the GSTs: this may be due to the relatively short half-life of the GST in the plasma.

3.2 GST in Alcoholic Cirrhosis

In alcoholic cirrhosis, abnormalities in the GSTs have been reported [27]. There is conflicting evidence in the literature on this point, the conflict possibly being based on the use of GST activity (rather than mass) measurements in some cases as well as on the use of different antisera having a range of specificities against the different forms of GST.

More recently with the use of well-defined antisera against the homodimeric forms of the basic liver GSTs (B_1B_1 and B_2B_2) it has become clear that the homodimers may be raised in the plasma in chronic alcoholic liver disease (Fig. 6). A higher proportion of patients showed increased B_1B_1 than B_2B_2. However, approximately 20% of the patients were found to have abnormal ALT activities with normal B_1B_1: a similar proportion had abnormal B_1B_1 values and normal ALT activities. The reasons for these discrepancies may be partly explained by the different distributions of the two enzymes in the liver (ALT is periportal and B_1B_1 centrilobular) and by differences in their biological half-lives.

4 Procollagen III Peptide and Hepatic Fibrosis

For a review of the biosynthesis and structure of collagen, and its disorders, refer to [28]. The three polypeptide chains of the collagen molecule are each coiled into a left-handed helix and in turn three of these helices are twisted around each other into a right-handed super helix, thereby forming a rigid structure about 1000-nm long. It is these rigid structures that are subsequently assembled into collagen fibrils.

Detailed chemical analyses have demonstrated that a variety of collagens are present in humans. All of these consist of the helices organised in general as described above. They also contain large amounts of proline, hydroxyproline and glycine, which represent about 50% of their total amino acid content. They differ, however, in the remaining amino acid content and are apparently the products of different genes. Type III collagen accounts for 10–50% of the total collagen content in adult tissues such as large arteries, lung, muscle and liver.

The interstitial collagens (I–III) are synthesised in the form of larger molecules, the procollagens, which contain additional precursor-specific peptide extensions (propeptides) at both the N-terminal and the C-terminal ends. During the conversion of procollagen into collagen these extensions are removed by specific proteinases, thus producing the native collagen fibrils. It appears that the presence of these propeptides, which increase the water solubility of the procollagens to which they are attached, enables the passage of the procollagen molecules from their site of synthesis to the site at which the insoluble collagen fibrils are to be laid down.

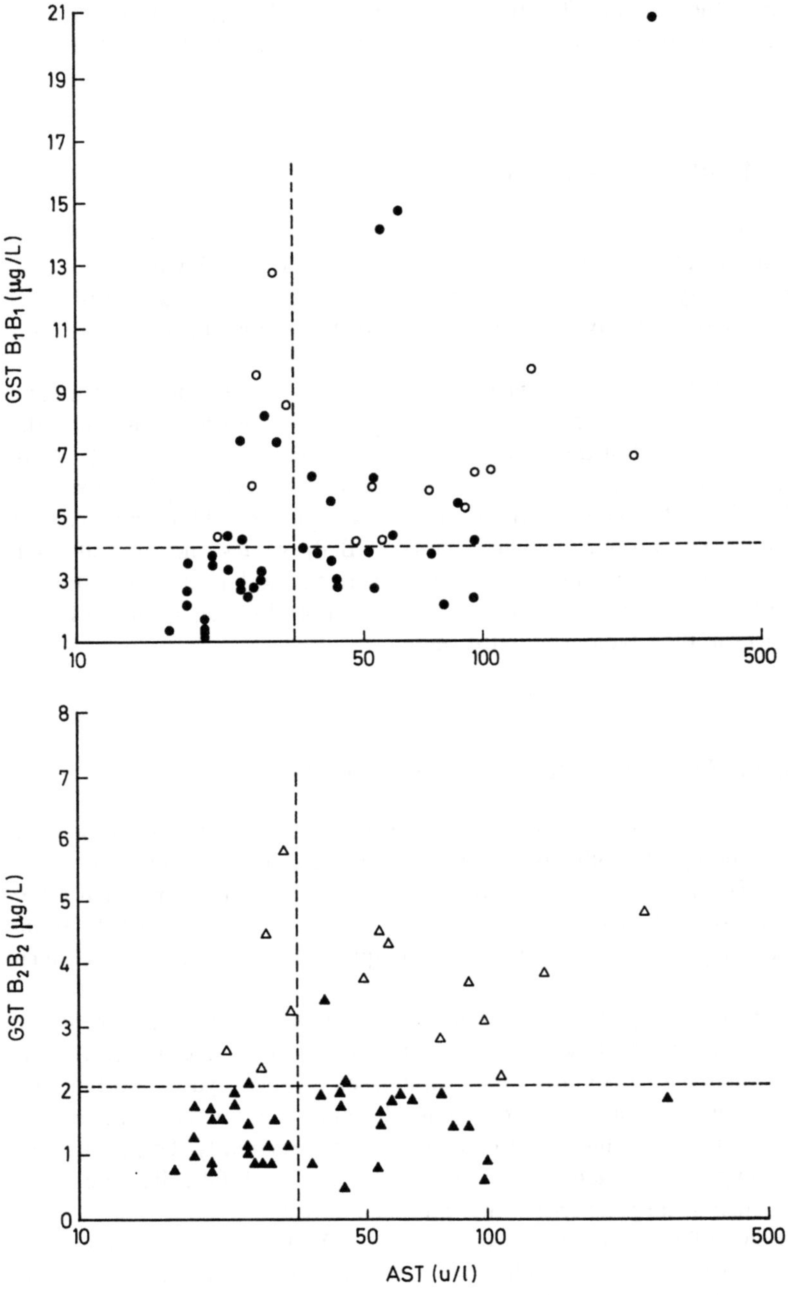

Fig. 6. The activity of aspartate aminotransferase (AST) and the concentration of the homodimers (B_1B_1 and B_2B_2) of glutathione-S-transferase (GST) were measured in 54 patients with alcoholic cirrhosis. Patients in whom both of the GST concentrations were abnormal are shown by (○) or (△). The upper limits of the reference intervals are shown by the dotted lines

The n-terminal propeptide from human type III collagen has been purified [29] and following bacterial collagenase digestion it has been shown to consist of an N-terminal non-collagenous domain in which cystine residues form intrachain disulphide bridges, a C-terminal non-helical domain and a collagenous domain. Antibodies raised against the human N-terminal propeptide react similarly with the purified human peptide and with a corresponding segment of bovine procollagen III.

Scars and fibrotic tissue consist primarily of collagen fibrils. Therefore, increased collagen deposition occurs in any disease associated with fibrosis. Since collagen deposition appears to involve the enzymatic cleavage of the procollagen molecule with the production of the propeptides, it seems logical to ask whether the process of laying down of fibrotic tissue can be monitored by measuring the presence of the propeptides and whether there is a quantitative relationship between the extent of fibrosis and the concentration of the propeptide.

The laying down of fibrotic tissue in the liver was extensively reviewed in 1975 [30]. It was shown that proline hydroxylase activity in liver biopsy tissue increases before there is any discernable increase in the collagen content of the liver and that proline hydroxylase activity in liver biopsy specimens may be useful in following patients with alcoholic liver disease [31]. In addition, immunoassay for proline hydroxylase in plasma showed that the amount of the enzyme rose in association with fibrosis of the liver [32]. The more recent purification of the N-terminal propeptide of collagen type III has enabled the analysis, by immunoassay, of the propeptide in liver disease.

Several measurements have been considered as non-invasive markers of hepatic fibrosis and hence as alternative methods by which the progress of patients with liver disease could be followed without a need for liver biopsy. These include urinary hydroxyproline levels [33] and serum proline levels [34]. Tanaka and his colleagues [35] examined these endpoints as well as measuring the collagen type III propeptide and compared the results with morphometric analysis of fibrosis in liver biopsy specimens from patients with alcoholic liver disease. They showed that in patients with hepatic fibrosis, serum proline levels and routine liver function tests were not significantly correlated to the area of fibrotic change measured by morphometry. However, the concentration of the procollagen-III-peptide showed a significant positive correlation with the area of fibrosis (Fig. 7). Hepatic prolyl hydroxylase activity in liver biopsy was also compared with the concentrations of procollagen-III-peptide in serum [36] in patients with alcoholic hepatitis. Both the enzyme activities and the peptide concentrations were significantly higher in patients with alcoholic hepatitis than in a control group. Both measurements were more abnormal in patients with severe fibrosis than in mild fibrosis and there was a close correlation between the activity of proline hydroxylase and procollagen-III-peptide concentration ($r = 0.76$, $p < 0.001$). It appears, therefore, that procollagen-III-peptide may constitute a good marker of hepatic fibrogenesis in alcoholic liver disease. Levels of the peptide appear also to reflect disease activity in autoimmune chronic active hepatitis [37]. The place of procollagen-III-peptide measurements in patients with primary biliary cirrhosis is however not clear. While the peptide appears to reflect the histological stage of the disease, its place as an independent predictor of survival of patients with primary biliary cirrhosis remains controversial.

Measurement of procollagen-III-peptide may be a useful index of fibrosis but it

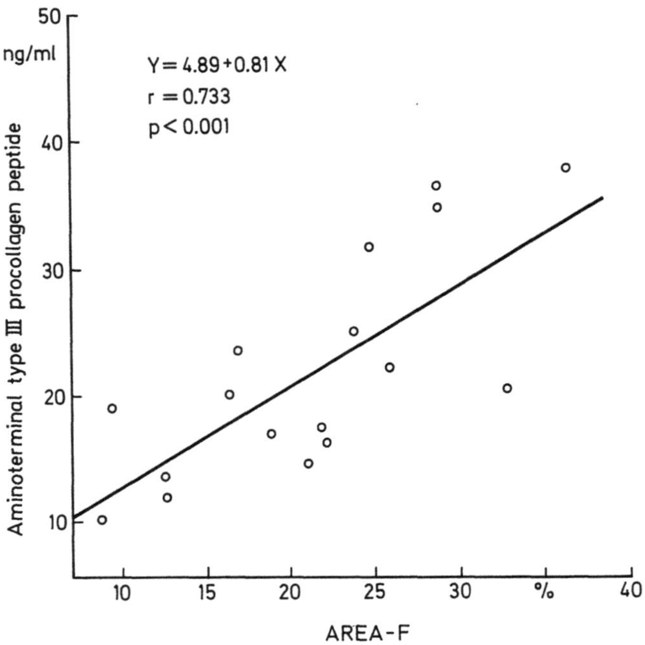

Fig. 7. The concentration of procollagen III peptide in serum was measured and the area of fibrosis in sections, prepared from liver biopsy tissue from patients with alcoholic liver disease, was measured using morphometric analysis

should be borne in mind that it is not specific to individual tissues and disease activity outwith the liver needs to be considered in patients with established liver disease. Thus, for example, increased levels of the peptide in serum have been reported in rheumatoid arthritis [38] and here too there appears to be a direct relationship between the disease activity and the peptide concentration. Increased levels of the peptide have been reported in a wide range of conditions including myelofibrosis [39], Paget's disease of bone [40] and malignant non-liver neoplasms [41]. Its place in the clinical biochemistry of the hepato-biliary system remains to be fully evaluated.

5 References

1. Carey MC (1982) in: Arias I, Popper H, Schacter D, Shafritz DA (eds). The liver: biology and pathobiology. Raven Press, New York
2. Hofmann AF, Molino G, Milanese M, Belforte G (1983) J Clin Invest 71: 1003
3. Molino G, Hofmann AF, Cravetto C, Belforte G, Bona B (1986) Eur J Clin Invest 16: 397
4. Ahlberg J, Angelin B, Bjorkhem I, Einarsson K (1977) Gastroenterology 73: 1377
5. Gilmore IT, Thompson RPH (1981) Clin Sci 60: 65
6. Mashige F, Imai K, Osuga T (1976) Clin Chim Acta 70: 79
7. Becket GJ, Hunter WM, Percy-Robb IW (1978) Clin Chim Acta 88: 257
8. Beckett GJ, Corrie JET and Percy-Robb IW (1979) Clin Chim Acta 93: 145
9. Roda A, Roda E, Aldini R et al (1977) Clin Chem 23: 2107
10. Karlaganis G, Paumgartner G (1979) Clin Chim Acta 92: 19

11. Mannes GA, Stellaard F, Paumgartner G (1987) Clin Chim Acta 162: 147
12. Percy-Robb IW (1985) In: Recent Advances in Clinical Biochemistry, CP Price, KGMM Alberti (eds) Churchill Livingstone pp 125
13. Beckett GJ, Dewhurst N, Finlayson NDC, Percy-Robb IW (1980) Gut 21: 734
14. Aldini R, Roda A, Festi D, Mazzella G, Morselli AM, Sama C, Roda E, Scopinaro N, Barbara L (1982) Gut 23: 829
15. Douglas JG, Beckett GJ, Nimmo IA, Finlayson NDC, Percy-Robb IW (1981) Gut 22: 141
16. Douglas JG, Beckett GJ, Nimmo IA, Finlayson NDC, Percy-Robb IW (1981) Eur J Clin Invest 11: 421
17. Roda A, Roda E, Sama C, Festi D, Aldini R, Morselli AM, Mazzella G, Barbara L (1982) Gastroenterol 82: 77
18. Owens D, Sherlock S (1973) Br Med J i: 559
19. Jansen PLM, Cuypers T, Peters WHM (1984) Eur J Clin Invest 14: 295
20. Gordon ER, Goreski CA (1982) Can J Biochem 60: 1050
21. Lauff JJ, Kasper ME, Ambrose RT (1981) J Chromatogr 226: 391
22. Smith GJ, Ohl VS, Litwack G (1977) Cancer Res 37: 8
23. Hayes JD, Strange RC, Percy-Robb IW (1980) Biochem J 185: 83
24. Hayes JD, Gilligan D, Chapman BJ, Beckett GJ (1983) Clin Chim Acta 134: 107
25. Beckett GJ, Hayes JD (1984) Clin Chim Acta 141: 267
26. Beckett GJ, Chapman BJ, Dyson EH, Hayes JD (1985) Gut 26: 26
27. Beckett GJ, Hayes PC, Hussey AJ, Bouchier IA, Hayes JD (1987) Clin Chim Acta 169: 85
28. Prockop DW, Kivirikko KI, Tuderman L, Guzman NA (1979a) NEJ Med 301: 13
Prockop et al (see above) (1979b) NEJ Med 301: 77
29. Niemela O, Ristelli L, Parkknen J, Ristelli J (1985) Biochem J 232: 145
30. Popper H, Becker K (eds) (1975) Collagen metabolism in the liver. Stratton Intercontinental Medical Book Corporation, New York
31. Mezey E, Potter JJ, Madrey WC (1976) Clin Chim Acta 68: 313
32. Tuderman L, Ristelli J, Miettinen TA et al (1977) EJ Clin Invest 7: 537
33. Prockop DJ, Kivirikko KI (1967) Ann Intern Med 66: 1243
34. Kershenobich D, Garcia-Tsao G, Saldana SA, Rojkind M (1981) Gastroenterology 80: 1012
35. Tanaka Y, Minato Y, Hasumura Y, Takenchi J (1986) Dig Dis and Sci 31: 712
36. Torres-Salinas M, Pares A, Caballeria J, Jimeney W, Heredia D, Bruguera M, Rodes J (1986) Gastroenterology 90: 1241
37. Eriksson S, Zethervall O (1986) J Hepat 2: 370
38. Harsleve-Petersen K, Bentsen KD, Junker P, Lorenzen IB (1986) Arth and Rheum 29: 592
39. Hochweiss S, Fruchtman S, Hahn EG, Gilbert H, Donovan PB, Johnson J, Goldberg JD, Berk B (1983) Amer J Hematol 15: 343
40. Simon LS, Krane SM, Wortman B, Krane IM, Kovity KL (1984) J Clin Endocrinol Metab 58: 110
41. Bolarin DM, Savolainen E-R, Kivirikko KI (1982) Int J Cancer 29: 401

The Serum Gamma-glutamyltransferase Isoenzyme System and its Diagnostic Role in Hepatobiliary Diseases

Lucia Sacchetti, Giuseppe Castaldo and Francesco Salvatore

Dipartimento di Biochimica e Biotecnologie Mediche, II Facoltà di Medicina, Università di Napoli, via S. Pansini 5, I-80131 Napoli, Italy

In this review some novel methodological and clinical aspects of human serum gamma-glutamyltransferase (GGT; E.C. 2.3.2.2) are presented and discussed within the framework of a general survey on the enzyme.

The various forms of the enzymes that have been found in tissues and blood serum differ from each other in several molecular features. Here the different forms are described with special reference to the serum forms whose biochemical properties are still poorly characterized. The molecular physiology of the enzyme is also briefly described with respect to the biochemical mechanisms in which the enzyme is involved.

Studies on serum GGT received an impetus from the introduction of a simple, rapid assay procedure that is endowed with a high sensitivity and reproducibility. With this improved methodology the GGT isoenzyme pattern in serum has been clinically correlated with various types of hepatobiliary diseases (persistent and active chronic hepatitis, cirrhosis, intra-hepatic and extra-hepatic obstructive jaundice, primary and metastatic liver neoplasia) and with acute pancreatitis. Satisfactory reproducible patterns are being obtained and may be used to contribute to the diagnosis and monitoring of several hepatobiliary diseases. Of special interest is the presence of a serum albumin-comigrating GGT band in primary and in metastatic liver tumors. The diagnostic specificity of this band is over 90% toward non-neoplastic liver diseases and toward non-hepatic tumors, whereas the diagnostic sensitivity gradually increases with the increase in the level of total serum GGT and with the stage of the neoplastic disease.

Several molecular properties of the multiple forms of serum GGT have been studied both in normal subjects and in patients affected by hepatobiliary diseases using sequential selective lipoprotein fraction precipitation. The GGT fraction complexed with low density lipoprotein plus very low density lipoprotein, at a certain cut-off level, gives rise to a diagnostic sensitivity of about 75% for liver tumor patients, and to a diagnostic specificity of about 80% as a discriminating power between liver tumor patients and those affected by cirrhosis and chronic hepatitis.

The perspectives for the clinical use of GGT serum isoenzyme forms thus appear more promising from these recent studies.

Progress in Clinical Biochemistry
and Medicine, Vol. 8

1 Introduction

Knowledge on the pathophysiology of the human serum gamma-glutamyltransferase, or transpeptidase (glutamine: D-glutamyl-peptide 5-glutamyl transferase, EC 2.3.2.2) isoenzyme system has been increasing in recent years. In particular, data have accumulated regarding its diagnostic role in hepatobiliary diseases. More than 25 years has passed since serum GGT activity was first indicated as a marker of liver disease [1]. Since that time several reviews have summarized the data obtained on biochemical and clinical aspects of this serum enzyme and its isoenzymes in health and disease [2-6]. However, it has not yet been possible to assign a definite, clear-cut role to this enzyme, particularly in regard to possible discriminating properties of these indicators toward the wide variety of hepatobiliary disorders [3-6].

Since the last review appeared in 1985 [6], additional data have been obtained in various laboratories where serum GGT is studied, including our laboratory. This series of studies has resulted in novel acquisitions and potentialities as far as the diagnostic role of this enzyme and its isoenzyme forms are concerned, and therefore it appears that an update of the data on these topics is appropriate at this time.

In this review, together with a brief description of the general properties and functions of human tissue and serum GGT, we shall provide a panorama of data on multiple forms of GGT, their estimation in blood, and the clinical correlation of the serum isoenzyme patterns with hepatobiliary diseases including neoplasias. Taken together these findings throw some light on this rather confused field, and provide indications for new advances concerning the pathophysiological aspects of this serum isoenzyme system.

Turning to the technical aspect, Nemesànszky and Lott [6] ended their exhaustive review by stating that "Methodological research should be promoted to standardize the procedural choice for clinical laboratories . . .". Also in this optics, recent data have changed the outlook, and have opened new possibilities that may be followed-up by clinical laboratories. Also the molecular nature of the various isoenzyme forms or isoforms present in serum has started to be unravelled even though we are still in a primitive stage of such studies.

2 Biochemical Properties and Functions of GGT

2.1 Mechanisms of the Reaction and Properties

Gamma-glutamyltransferase or transpeptidase (glutamine: D-glutamyl-peptide 5-glutamyltransferase, EC 2.3.2.2) is a membrane glycoprotein [7] that transfers the gamma-glutamyl moiety of glutathione to a variety of acceptors, as shown in the scheme below [8]. In most cases the acceptor is an amino acid or a second molecule of glutathione or a water molecule, thus releasing in all cases a cysteinyl-glycine dipeptide. In the third case, the nucleophyl group is the water molecule and the reaction determines the hydrolysis of glutathione into glutamate and the already mentioned dipeptide.

Gamma-glutamyltranspeptidase reaction mechanisms

1) Glutathione + amino acid $\underset{}{\overset{GGT}{\rightleftarrows}}$ gamma-glu-amino acid + CYS-GLY

2) 2 Glutathione $\overset{GGT}{\rightleftarrows}$ gamma-glu-glutathione + CYS-GLY

3) Glutathione + H_2O $\xrightarrow{GGT}$ glutamate + CYS-GLY

Apart from the mechanisms of the enzymatic reaction catalyzed by the enzyme, its overall physiological role is to contribute to the intracellular transfer of the amino acids. This has been widely described and illustrated by Meister [9] in the overall glutathione cycle, or rather gamma-glutamyl cycle, which results in the degradation and the biosynthesis of glutathione, while amino acids are translocated through the membrane into the cell (see Fig. 1).

2.2 Occurrence in Human Cells and Body Fluids

2.2.1 Tissues

Even though these biochemical functions are, of course, unrelated to the role of GGT as a biochemical indicator of disease, they do illustrate that GGT isoforms are widely distributed in human tissues with the greatest activity occurring in cells and tissues with high secretory activity or with absorption functions [10-18], i.e. the epithelial cells of the proximal renal tube [13,14,18], of the jejeunal intestine [10], of the biliary tract, of the hepatocytes [12], at the internal cannular surface of the prostate gland [16], of the small bronchus [6], and of thyroid follicle cells [6]. High concentrations of the enzyme have been described also in the pancreas [15,17]; mammary gland also produces a high amount of GGT [11], elevated concentrations of which appear in the milk [19,20].

Several studies have been performed on partially purified or purified GGT forms isolated from human tissues. The enzyme produced by human kidney is a glycoprotein with a M_r of 84,000 composed (as are all the forms of GGT) of two glycopeptides that have a M_r of 62,000 and 22,000, respectively [21]. Pancreas [15,17] produces two different forms of GGT that have a different rate of glycosylation; they could derive from different pancreatic cell populations, such as ductal cells and acinal cells; their M_r is identical. It is 88,000 [15]: 61,000 and 27,000, respectively, for the two subunits; others have obtained a M_r of 110,000 [17].

GGT from liver has been widely studied. Huseby [22] reported a M_r for the enzyme of 90,000 obtained with electrophoresis and 110,000 with gel filtration. He also found that the liver and serum enzymes had similar kinetic properties. Shaw et al. [23] reported that the liver and kidney enzymes had similar kinetic properties, although they found a great difference in the degree of inhibition by glycylglycine. Shaw et al. [24] also reported a different electrophoretic migration of GGT from liver, pancreas, kidney, and intestine on acrylamide gel, and a widely different inhibition constant for these multiple forms by glycylglycine. The calculated M_r of the tissue GGT isoforms is reported in Table 1.

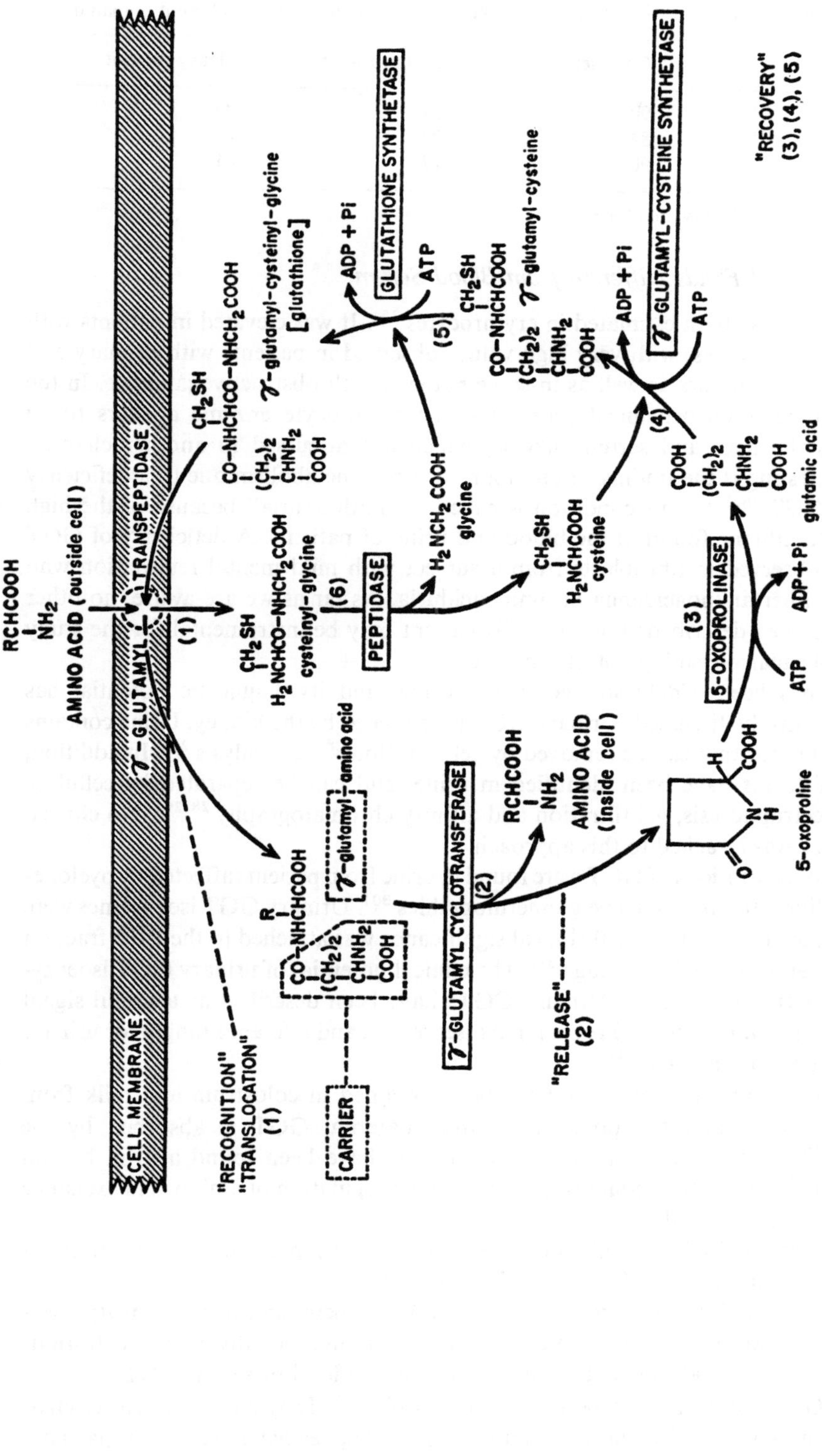

Fig. 1. The γ-glutamyl-cycle (reproduced with permission from: Meister A, 1973 *Science*, p 35)

Table 1. Molecular weight ($\times 10^3$) of gamma-glutamyltransferase forms from different human tissues

		Complete	Light subunit	Heavy subunit
Liver	22)	90	22[a]	47[a]
Kidney	21)	84	22	62
Pancreas	15)	90	27	61

[a] measured on non-glycosylated forms

2.2.2 *Biological Fluids Different from Blood Serum*

GGT activity has been estimated in erythrocytes [25]. It was elevated in patients with hepatobiliary diseases, with very high values observed in patients with primary and metastatic liver tumors as well as in some patients with obstructive jaundice. In the light of some physicochemical properties, the erythrocyte enzyme appears to be similar to the liver and serum enzyme, which are discussed in another chapter. Quite interesting is the finding of an inborn error of metabolism due to a deficiency of tissue GGT [26,27]. This condition is called "glutathionuria" because of the high level of glutathione found in the blood and urine of patients. A deficiency of GGT has been observed in fibroblasts from a subject with mild mental retardation who showed neither aminoacidemia or aminoaciduria. As far as we are aware, no other cases have been described; however, this patient may be instrumental for the study of the molecular physiology of the enzyme.

GGT has been widely studied in the urine, and its diagnostic potential has been exhaustively discussed. Urinary GGT is produced by the kidney. Urine contains a GGT inhibitor that can be removed by gel filtration [28], or dialysis [29]. In addition, two GGT forms have been identified in urine, and can be separated by cellulose acetate electrophoresis, gel filtration and affinity chromatography [28,29]. No clinical significance was attached to this approach.

High concentrations of GGT were found in urine from patients affected by pyelonephritis, Wilms' tumor and some glomerulopathies [30]. Urinary GGT isoenyzmes were further studied, and a potential clinical significance was attached to the P105 fraction as an indicator of tubular damage [31]. The clinical potential of urinary GGT isoenzymes was later confirmed [32]. Urinary GGTs have been described as a useful signal in monitoring patients with chronic renal diseases [33], and reference ranges for urinary GGT have been evaluated [34].

High concentrations of GGT have been obtained in colostrum and milk from goats and foals, and the postulate is that maternal GGT is absorbed by the newborn [19]. High concentrations of the enzyme have been found also in human colostrum [20], where they could be involved in the regulation of amino acid exchange between tissues and milk.

A decrease of GGT and alkaline phosphatase (ALP) in bronchial aspirate could be a useful signal for bronchogenic malignancy [35].

Studies on GGT in saliva revealed no correlations between age and salivary levels of GGT or between the enzyme and sex [36]. Furthermore, the salivary concentrations of GGT were also independent of fluctuations in the level of serum GGT.

Four GGT isoforms have been identified in bile [37]. They have different electrophoretic mobility, molecular size, and density. The enzyme has been partially purified in bile, where it appears with a concentration 150-fold higher than in serum [38]. The origin of the biliary enzyme, and its relationship with the serum and liver enzymes, particularly in patients with biliary obstruction is still obscure [39].

2.3 *Biosynthesis*

Recent *in situ* hybridization experiments on human chromosomes confirmed, by using a rat cDNA specific for this enzyme, that one gene is present for the haploid genome on human chromosome 22 at the interface of q111–112, and a minor peak on q131 [40]. Other studies performed on cDNA from rat kidney indicated the presence of one mRNA precursor for the two polypeptide chains by which the enzyme is formed [41,42]. In fact, the enzyme protein molecule is always constituted by two different monomers of very different M_r, a light chain of M_r around 25,000 and a heavy chain about three-fold higher. They derive by proteolytic action from a unique precursor proenzyme. Both monomers contain a high proportion of carbohydrate. The heavy subunit includes the hydrophobic sequence corresponding to amino acids 1–21 or 6–24, which is considered to be the leader sequence by which the enzyme is attached to the membrane [43]. The light subunit is the one that contains the active site of the enzyme [44]. A high degree of similarity is found between the rat and human small subunit: both contain the active site, and they show a similar M_r and amino acid composition [21]. Marked differences in M_r and amino acid composition have been found between the two large subunits. The large subunit could contain a further "latent" active site. The isolated large subunit of the rat kidney enzyme catalyzes the enzymatic decomposition of the substrate γ-glutamyl-*p*-nitroaniline with the same Km values and pH-activity curve as the native enzyme [45]. This active site is inoperative in the native enzyme molecule. The complete GGT sequence obtained by the analysis of this cDNA confirms that the two GGT subunits originate from a common precursor [42, 46, a].

Many studies have been devoted to the biosynthesis of rat renal GGT. Both in vitro [47,48] and in vivo [49] studies have clarified the sequence of events that occur in GGT biosynthesis. A single polypeptide, comprising a hydrophobic leader peptide, with a M_r of 78,000 [47] is synthesized. The cleavage of the enzyme occurs during the cytoplasmatic passage of the enzyme. The propeptide cleavage occurs at two sites [50]. The two subunits have a M_r of 50,000 and 23,000, respectively. Glycosylation of the enzyme occurs in the Golgi complex, where the propeptide is bound to the "core" and peripheral sugars [49] before its cleavage. Membrane insertion of the subunits (bound to each other with non-covalent bonds), via the NH_2 terminal hydrophobic sequence [51], is the last event of GGT biosynthesis. Also the immunogenicity of the various GGT forms isolated from different tissues indicates that they are very similar. Indeed, all forms crossreact with human liver antiserum [52, 53]; the rat and human

---

[a] While this article was being printed, a paper appeared [112] that described the isolation of GGT genome sequences from rat and human libraries. At least four genes for GGT were demonstrated; however it is not known if these genes are tissue-specific, if they are peculiar to diverse ontogenic phases, or if they are pseudogenes. In addition, since the present work was written, Rajpert and co-workers [113] have cloned and sequenced human GGT; they found that the human and rat nucleotide sequences differ greatly in a 180-base pair fragment of the heavy chain — a discrepancy that results in a different amino acid sequence.

kidney enzymes also crossreact, and it has been concluded that the two forms have some antigenic determinants in common [52].

Whereas the isoforms have the same amino acid sequence because they derive from the same gene, they differ from each other in post-translational modifications, particularly glycosylation [44], which gives rise to various types of side-chains harbouring ramifications of these carbohydrate moieties. The site of attachment is the asparagine moieties of the polypeptide chain residues present in the ASN-X-SER sequence. The problem of the release of the enzyme within the bloodstream is rather complicated, because the isoforms that have been found in the bloodstream are entirely different from those of tissues with respect to molecular weight and to chemical composition. Indeed, the enzyme binds to several serum components, mainly macromolecular components, such as lipoproteins [55–60], other proteins [61,62], membrane fragments [63] and also lipids. The post-translational modifications that have been found in pathological situations, particularly tumors, are reported in Sect. 4 and 5 below.

3 GGT Multiple Forms and their Estimation in Blood

3.1 Normal Individuals and Different Pathophysiological Conditions

The possible mechanisms by which the enzyme is released from the various cells present in liver tissue into the bloodstream are schematically illustrated in Fig. 2 [64]. Any type of liver cell, the parenchymal, the biliary epithelium, or Kupffer cells may

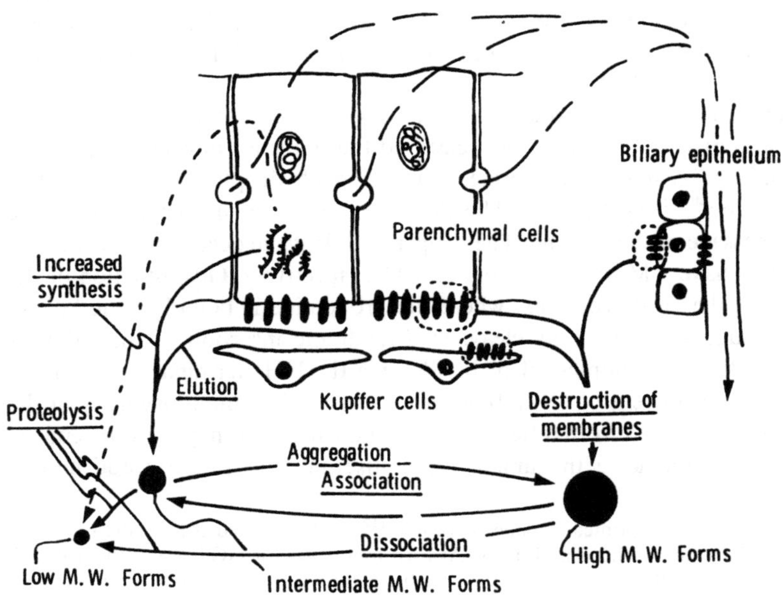

Fig. 2. Possible derivation of serum GGT isoforms from hepatobiliary cells. (Reproduced with permission from: Moss DW, 1984 in: Goldberg DM, Werner M, eds., Selected Topics in Clinical Enzymology, vol 2, p 525)

release the enzyme from the membrane upon destruction or derangement. Increased synthesis may cause the enzyme to delay its attachment to the membrane.

The types of enzymes present in serum depend on the kind and the extent of cell alteration. Therefore, the logical step was to try to relate the presence and the type of the serum GGT forms to the nature of the cell alteration, and, what is more important for diagnostic purposes, to the nature of the liver damage.

Looking first at the molecular weight, three main classes have been identified [65-67]. The high molecular class forms, which have a M_r of over one million, include the enzyme attached to cell membrane fragments, and association with the same and other types of enzyme molecules. These forms, described "hydrophobic", are abundant in sera from patients with cholestasis [55]. They could be due to an increased release of the enzyme from the canalicular membrane, probably due to the detergent effect of biliary salts [57,65-67], and to the further complex of the enzyme (via the short hydrophobic polypeptide) with kilomicrons and/or low density lipoproteins (LDL) and very low density proteins (VLDL) [55-59], and particularly with lipoprotein X [55]. It could also result from a complex of the enzyme with membrane fragments that have been described in sera from patients with severe hepatobiliary jaundice [63]. The intermediate M_r class, corresponding to 250–450,000, could derive from a complex between the hydrophobic enzyme and high density lipoproteins (HDL) [55-59]. These groups of complexes have been described in sera from patients with non cholestatic liver damage [55-62]. The third low M_r type, corresponding to 80–120,000, derives from the previous forms directly by overflow, often after administration of alcohol and enzyme-inducing drugs [68-71]. Some authors consider it a typical form of normal subjects [57,58], but others have not found it in normal subjects [48-50]. The synthesis and hence the serum concentration of this isoform could be increased after reticular induction. Alcohol [68,69] and drugs such as barbiturates [70,71] could cause this induction. Low-molecular-weight GGT may also be present in some cases after proteolysis, or perhaps, incomplete synthesis of the enzyme molecule [57]. The liver is generally held to be the main source of serum GGT in normal individuals, and also in situations of organ and tissue alteration. In fact, it seems that kidney, pancreas, and intestine, which show the highest concentrations of the enzyme, contribute very little, if at all, to serum enzyme concentration (see [6]). Further modifications of the GGT molecule have been described in neoplastic diseases and these will be discussed in detail Sect. 5, below.

Taken overall, the results of studies on the multiple molecular forms of serum GGT have generated more confusion than they have answered questions. Since the data been obtained with different methodologies (see [6,72]) and since different nomenclatures have been adopted for the electrophoresis bands, the findings are often not comparable, and not sufficiently consistent. In fact, serum GGT isoenzymes have been evaluated with electrophoresis on agar [73], cellulose acetate [70,74-79] and agarose [80,81], and different substrates [82] have been used for band visualization. Electrophoresis on polyacrylamide gels has revealed some serum GGT isoforms peculiar to cancer [83-85]. With this method it is possible to discriminate between intrahepatic and extrahepatic cholestasis [56]. Recently, also isoelectric focusing has been used to evaluate serum multiple forms of GGT [86-88]. With this method, isoforms of GGT with a different glycosylation rate can be easily identified; but no findings of clinical relevance have resulted from this separation. In addition to electrophoresis, different types of affinity chromatography [89,90] and gel filtration [91] have been used to study the GGT isoforms in sera from normals and hepatobiliary subjects. Very recently an HPLC separation method has been described; however [92], the process is rather complex because of the frequent association of GGT with other types of small molecules.

Since GGT is a glycoprotein with a high percentage of carbohydrate moieties, the variety and abundance of oligosaccharide chains covalently bound to the protein backbone determines a variety of possible forms, which contribute to complicate the isoenzymatic serum pattern of these proteins.

3.2 *Methodology for GGT Isoform Separation in Blood*

Given the findings discussed above, it is obvious that the separation and estimation in blood of the various isoforms of serum GGT is of paramount importance. It became essential to have a method, easily reproduced in the clinical laboratory that would allow us to obtain data on larger populations than hitherto studied and to make quick and reproducible comparison with data obtained in other laboratories. Therefore, we devised an improved method [93] to comply with these requirements, and here we shall briefly summarize the main methodological steps as well as the advantages of its use in the clinical laboratory. Table 2 lists the main features of our electrophoretic method as compared with other commonly used procedures.

The buffer is obtained by dissolving 18 g of HR buffer powder (Helena Lab., Beaumont Texas) for the electrophoretic migrations in 750 ml (pH = 8.4) of distilled water. Electrophoresis is performed on cellulose acetate plates for lipoprotein separation (Titan III Lipo, Helena) for 40 min at 220 volts at room temperature. After electrophoresis, the acetate cellulose strip is sandwiched between a second cellulose acetate supporting strip containing 1 ml of the visualization reagent. The latter is made by dissolving by sonication 3 mg of γ-glutamyl-7-amino-4 methylcoumarine (Cat. no. 7261, Sigma Chemical Co, St. Louis, MO) in 1 ml of methoxyethanol and then adding 400 μl of this to 10 ml of substrate solution containing 242 mg of Tris base and 132 mg of glycylglycine (Sigma Chemical).

The sandwich is then opened and the serum-containing strip dried (10 min at 60 °C) and submitted to U/V densitometer scanning. With the use of our electrophoretic method, an analytical sensitivity of about 10 U/L of total GGT (at 25 °C) and about 18 U/L (at 37 °C) have been obtained. With this method we identified eight different isoforms of GGT in different hepatobiliary diseases, and two isoforms typical of normal individuals; their nomenclature is based on normal protein electrophoresis.

Table 3 summarizes from the literature the various molecular and physico-chemical characteristics of the serum GGT forms. The table includes our recent data, in the sense that the enzyme forms are listed according to their electrophoretic migration on a cellulose acetate support and the nomenclature corresponds to conventional electrophoretic migration of serum proteins. The GGT form that showed no mobility is referred to as dep-GGT. The table also shows the possible correspondence with isoforms described by other authors obtained with different methodologies, and their probable M_r based on the same comparison. Some physico-chemical characteristics and chemical compositions are also indicated together with

Table 2. Methods used for electrophoretic separation of serum gamma-glutamyltransferase isoenzymes

	(93)	(76)	Helena	(79)
Buffer	pH 8.6 Tris-sodium barbital (HR from Helena 18 g/750 ml)	pH 8.8 Tris-sodium barbital (HR from Beckman)	pH 8.6 Tris-sodium barbital (HR from Helena 18 g/1500 ml)	pH 8.4 barbital buffer (B_1 from Helena 2.1 g/750 ml)
Support	cellulose acetate for lipoprotein (Titan III Lipo Helena)	cellulose acetate (Beckman)	cellulose acetate for isoenzymes (Titan III Iso Vis Helena)	cellulose acetate for isoenzymes (Titan III Iso Vis Helena)
Migration (Voltage, time)	220 V, 40 mins	250 V, 45 mins	180 V, 25 mins	180 V, 25 mins
Identification (Compound)	a	a	b	c
Visualization + scanning	Fluorescence (UV Scanning)	Fluorescence (UV Scanning)	Colorimetry 520 nm	Colorimetry 610 nm

a: gamma-glutamyl-7amino-4methylcoumarin
b: gamma-glutamyl-*p*-nitroaniline + 7.5% trichloroacetic acid (TCA)
c: gamma-glutamyl-3,5-dibromo-4-hydroxylanilide

Table 3. Main features of the serum gamma-glutamyltransferase isoforms

Cellulose acetate mobility	Prevalent occurrence in	Possible composition	Presumed Mr ($\times 10^3$)	Physical structure
$alpha_1$	a) Normals	GGT with no lipoprotein	90–120[a]	Hydrophilic
	b) Cirrhosis and non cholestatic hepatobiliary diseases	GGT + HDL	250–500	Hydrophobic
$alpha_2$	Normals	GGT with no lipoprotein	90–120[a]	Hydrophilic
beta	Cholestatic hepatobiliary diseases; pancreatitis	GGT + LDL-VLDL-lipoprotein X	600–1,000	Hydrophobic
gamma-dep	Cholestatic diseases (particularly extrahepatic)	GGT + LDL-VLDL-lipoprotein X	600–1,000	Hydrophobic
albumin	Liver tumors	GGT with no lipoprotein		Hydrophilic
beta-gamma pre-beta	Hepatobiliary malignancies	GGT + LDL-VLDL-LDL-VLDL-lipoprotein X		Hydrophobic

[a] The different electrophoretic mobility is probably due to a different glycosylation rate

correlation with hepatobiliary diseases. The latter will be discussed in the following section.

The imprecision (CV) of our electrophoretic procedure was evaluated by densitometric scans of bands in sera from normal subjects and in different hepatobiliary conditions, in order to test all the different fractions. The results (Table 4) were very satisfactory: for all isoenzyme fractions, the CVs ranged from 1.35% to 5.26% (within day) and from 4% to 6.17% (between days). The imprecision of repeated scans of the same samples, resetting the densitometer before each scan, was always <5% (range 1.10% to 4.10%).

Using the described procedure, the GGT isoenzymatic pattern was highly reproducible in 142 normal individuals [94] (see Fig. 3). All the subjects showed only two GGT fractions, $alpha_1$-GGT and $alpha_2$-GGT. The concentration of $alpha_1$-GGT was always from 1.5 to 4 times higher than $alpha_2$-GGT, as determined with an ultraviolet densitometer. The distribution diagrams of the $alpha_1$-GGT and $alpha_2$-GGT percentage values and the relative skewness and kurtosis indexes are shown in Fig. 4. Both distributions were gaussian. The reference intervals obtained by both methods, i.e. parametric (mean ± 2 SD: $alpha_1$-GGT: 57.4 to 80.7% and $alpha_2$-GGT: 18.5 to 41.5%) and non-parametric (97.5—2.5 percentiles: $alpha_1$-GGT: 59.5 to 81.5% and $alpha_2$-GGT: 19.2 to 42.4%) almost completely overlap. Our pattern was in agreement with the normal pattern reported by other groups. Other authors [75] described two forms of GGT in sera from normal individuals with an electrophoretic

Table 4. Imprecision of the method for gamma-glutamyltransferase isoform separation by Sacchetti et al. [93]. Isoenzyme bands in serum of normal neoplastic subjects. (Reproduced with permission from: Sacchetti L, Castaldo G, Fortunato G, Salvatore F (1988) Clin Chem 34: 420)

Subjects	Percentage of total GGT (average value after several determinations)		Imprecision as CV %		Between day
			Within scan (n = 10)	Within day[a] (n = 6)	(n = 10)
Normal					
(GGT = 25 U/L)	α_1	60.58	2.75	3.29	4.00
	α_2	39.48	4.10	5.26	6.00
Neoplastic					
(GGT = 275 U/L)	Alb	40.65	1.10	2.94	4.00
	α_1	25.73	1.70	1.81	4.80
	β	25.45	1.35	2.48	6.17
	γ	8.08	3.86	2.18	4.88

[a] Within day data are obtained also within the same plate (made of 6 slots); therefore, the low CV obtained is because the densitometer was not reset.

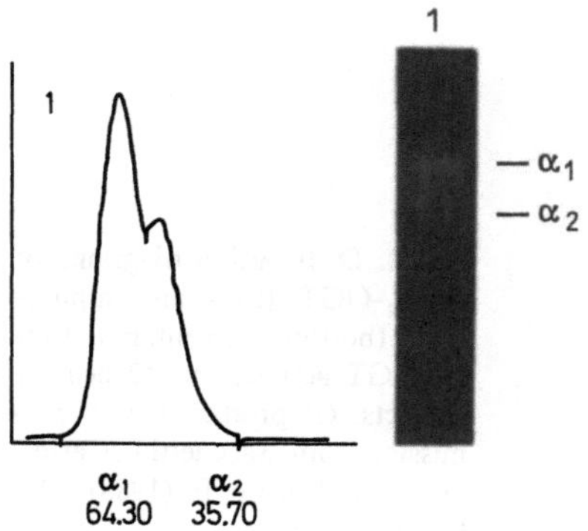

Fig. 3. Typical GGT isoenzyme pattern of a normal subject

mobility of alpha$_1$- and alpha$_2$-globulin, but they did not indicate the relative percentage. Rosalki [95] described in a third of his normal individuals another GGT fraction comigrating between the alpha$_1$- and alpha$_2$-fractions; he suggested that this band was related to an excessive alcohol intake. However, in other studies the alpha$_2$-GGT fraction (GGT co-migrating with alpha$_2$ serum proteins) was found to be the most abundant in most normal individuals [80], while under the experimental conditions of our method the most abundant form was alpha$_1$-GGT (see Fig. 3). This discrepancy could be due to our use of a different cellulose acetate support. The possibility that the two isoenzyme fractions are exactly inverted in their migration under the experimental conditions shown in Fig. 3 is reinforced by the fact that the two respective proportions are exactly reversed with respect to the values obtained when using another cellulose acetate support (Fig. 5). Furthermore, when only a single GGT band was present in the serum of a patient affected by cirrhosis, this band comigrated with alpha$_1$-GGT under the conditions of our method, but migrated with

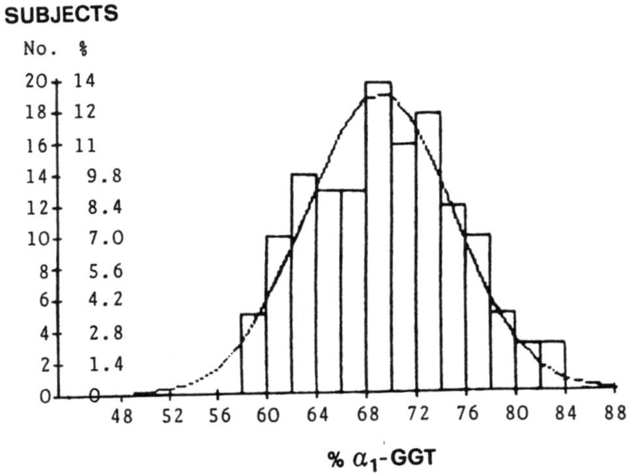

KURTOSIS INDEX 2.36

SKEWNESS INDEX +0.17

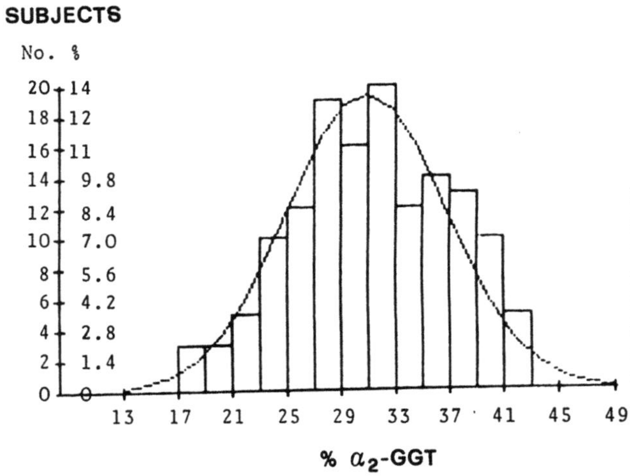

KURTOSIS INDEX 2.36

SKEWNESS INDEX -0.14

Fig. 4. Distribution diagrams of alpha$_1$-GGT (top) and alpha$_2$-GGT (bottom) percentages of total GGT activity in 142 healthy subjects. (Reproduced with permission from: Sacchetti L, Castaldo G, Salvatore F (1988) Clin Chem 34: 353)

alpha$_2$-GGT on the electrophoresis strips usually recommended for isoenzyme separation (Fig. 6). This conclusively demonstrates that alpha$_1$ and alpha$_2$ bands invert their migration when different electrophoretic acetate cellulose supports are used: conventional isoenzyme strips as compared with lipoprotein strips.

The normal subjects used for reference interval assessements were subdivided into classes according to age and sex. No significant variations were detected after statistical analysis, as shown in Table 5.

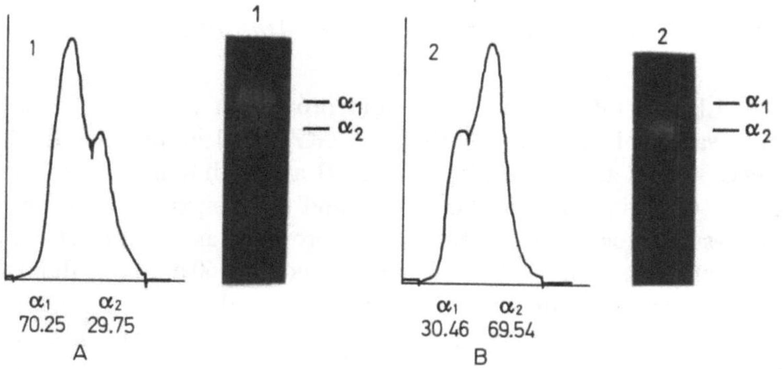

Fig. 5. GGT isoenzyme patterns obtained from the serum of normal subjects by the described method [93] with Titan III lipo cellulose acetate support (Helena Lab) (panel A), and with Titan III Iso-Vis cellulose acetate support (Helena Lab) (panel B). (Reproduced with permission from: Sacchetti L, Castaldo G, Fortunato G, Salvatore F (1988) Clin Chem 34: 421)

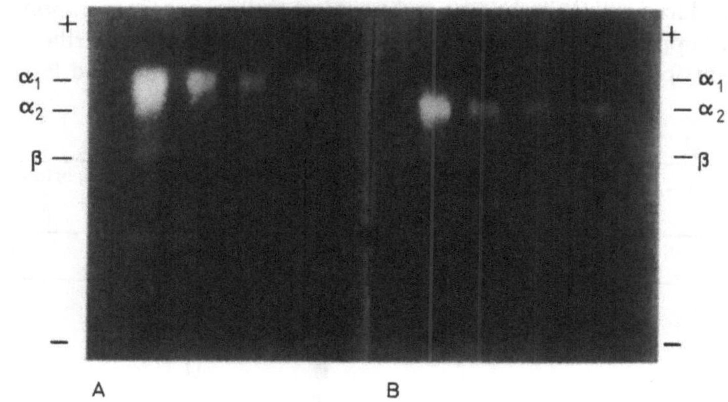

Fig. 6. Effect of support medium on electrophoretic patterns of GGT isoenzymes in serum from a cirrhotic subject: A, strip for lipoprotein measurement (Titan III Lipo, Helena Lab.); B, strip usually used for GGT isoenzyme measurement (Titan III Iso-Vis, Helena Lab.) (Reproduced with permission from: Sacchetti L, Castaldo G, Salvatore F (1988) Clin Chem 34: 355)

Table 5. Breakdown of population studied for gamma-glutamyltransferase isoenzyme reference intervals

Age		N	Alpha$_1$-GGT % ($\bar{x} \pm$ SD)	Alpha$_2$-GGT % ($\bar{x} \pm$ SD)
18–30	Males	17	69.3 ± 4.6	30.6 ± 4.6
	Females	17	67.5 ± 5.3	32.5 ± 5.3
31–40	Males	16	68.0 ± 4.4	32.0 ± 4.4
	Females	10	71.8 ± 6.1	28.2 ± 6.1
41–50	Males	18	72.0 ± 4.9	28.0 ± 4.9
	Females	18	70.5 ± 7.0	29.5 ± 7.0
51–60	Males	16	67.4 ± 6.6	32.6 ± 6.6
	Females	13	65.1 ± 4.5	34.9 ± 4.5
>60	Males	11	68.3 ± 7.1	31.7 ± 7.1
	Females	13	70.6 ± 5.7	29.4 ± 5.7

3.3 GGT in Fetal and Maternal Serum and in Amniotic Fluid

The ratio between ALP and GGT in amniotic fluid provides a measure of fetal pulmonary maturity (values of the ratio >2.0) [96]. A GGT level in amniotic fluid 20% below confidence limit is associated with trisomy 21 and with fetus affected by neural tube defects [97]. GGT in amniotic fluid was found to be significantly below the normal range in cystic fibrosis, and therefore it was proposed as a useful test for the prenatal diagnosis of this disease [98]. In an extensive study of 360 amniotic fluids, GGT fluctuation was associated with some important fetal alterations [99]. A comparative study of GGT and its isoenzymes in maternal and fetal serum and in amniotic fluid revealed that the latter contains at least two isoenzymes, the major form of which shows the same electrophoretic mobility as the fetal isoenzyme. In all cases, while mother and fetal enzyme activity remained constant, there was a significant decrease of GGT in amniotic fluid during gestation [100].

GGT activity has been also studied in cord sera [101]. It has been postulated that cord serum GGT is produced mainly in the liver and therefore the higher concentration found in cord serum as opposed to adult serum is due to hepatic immaturity. Furthermore, the predominant form of the serum GGT is complexed with HDL and less

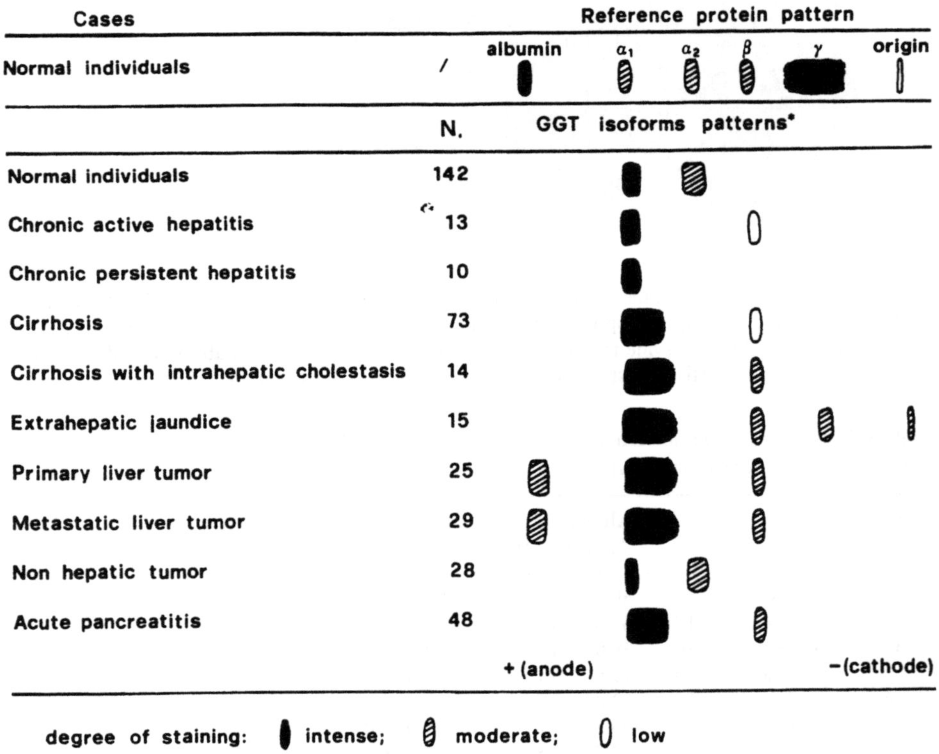

Fig. 7. Serum GGT isoform patterns, as obtained in normal individuals and affected by hepatobiliary diseases after the methodological approach described in Sect. 3.2. The patterns exemplified here are the most common ones found in the examined pathophysiological conditions (see text for variations)

sialized. We shall discuss the nature of the serum GGT isoenzyme forms in Sect. 6 of this article.

4 Clinical Correlations Between Serum GGT Isoenzyme Patterns and Hepatobiliary Diseases

The review by Artur and co-workers [72] and that by Nemesànszky and Lott [6] have described most of the studies concerning correlations of GGT isoenzyme pattern in serum with diseases of a hepatobiliary nature. Here we shall describe our studies, performed using the improved methodology described in the previous section, of a variety of pathological situations (see Table 3). The isoenzyme patterns obtained were highly reproducible within the groups of patients affected by the same type of disease, as shown in Table 3 and in Fig. 7.

The patterns typical of each type of disease are shown in Figs. 8, 9 and 10. Alpha$_1$-GGT and a small fraction (<10% of the total GGT) of beta-GGT were found in active chronic hepatitis and cirrhosis. This result is in agreement with the data published by Huseby [55], who indicated a prevalence of a GGT form migrating with

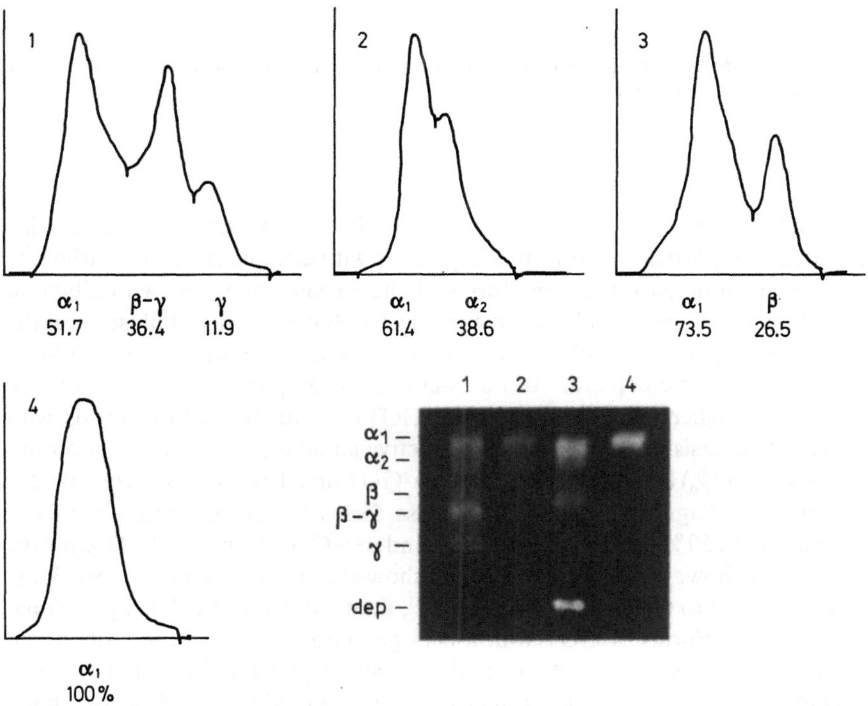

Fig. 8. GGT isoenzymes in different hepatobiliary diseases. 1: cholangiocarcinoma; 2: normal; 3: obstructive jaundice; 4: chronic hepatitis

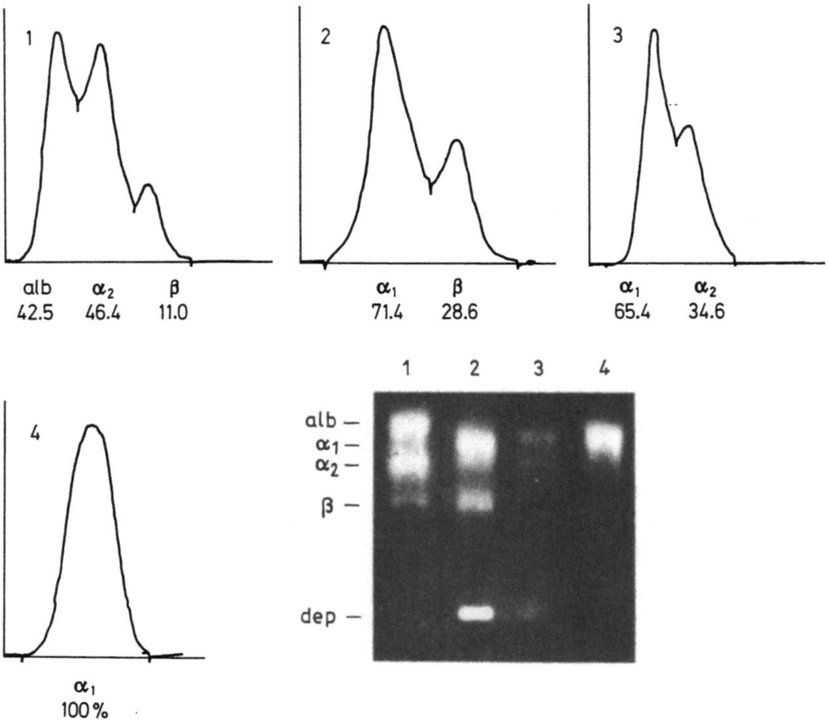

Fig. 9. GGT isoenzyme patterns in different hepatobiliary diseases. 1: liver neoplasia; 2: obstructive jaundice; 3: normal; 4: chronic hepatitis

alpha$_1$-globulin on agarose in sera from hepatobiliary, non icteric patients; also Hetland et al. [80] reported similar results in patients with alcholic cirrhosis. Burlina [75] described, in cirrhotic patients, some forms of the enzyme migrating with albumin, alpha$_1$-globulin and beta globulin using electrophoresis on cellulose acetate. Rosalki [95] also reported a predominance of GGT migrating with alpha$_1$-globulin in cirrhosis and chronic hepatitis. In our studies, the proportion of the beta-GGT form was much higher ($\leq 30\%$ of the total GGT) in patients with cirrhosis with intrahepatic cholestasis. In extrahepatic obstructive jaundice, in addition to a prominent (average 10–30%) beta-GGT band, gamma-GGT and dep-GGT isoenzymes were also present. The diagnostic sensitivity of these forms for extrahepatic obstructive jaundice was about 50% for the gamma-GGT and dep-GGT forms, and 100% for the beta-GGT band; however the first two forms showed a higher diagnostic specificity, evaluated according to Galen and Gambino [102], than did the beta-GGT (approximately 80% vs 10%). Forms of GGT with a beta-globulin electrophoretic mobility on agarose have been described as typical of cholestatic patients [55], and they result from a complex of GGT with lipoprotein-X (see below). Alpha$_2$- and beta-globulin co-migrating GGT with electrophoresis on agarose gel has been described as being typical of malignant diseases [80]. An increase of beta-globulin comigrating GGT in sera from cholestatic patients has also been reported by Burlina [75] and Rosalki [95]. Table 6 shows a detailed association that we found between various diseases and GGT isoenzyme patterns, showing the number and frequency of the single GGT isoenzyme bands in relation to the various hepatobiliary diseases examined. In several instances, particuarly in extrahepatic obstructive jaundice and liver tumor, the alpha$_2$-GGT band appears evident, and well distinct from the alpha$_1$-band (see Table 6), whereas in most other instances a large alpha$_1$-fraction covers the position where alpha$_2$-GGT migrates.

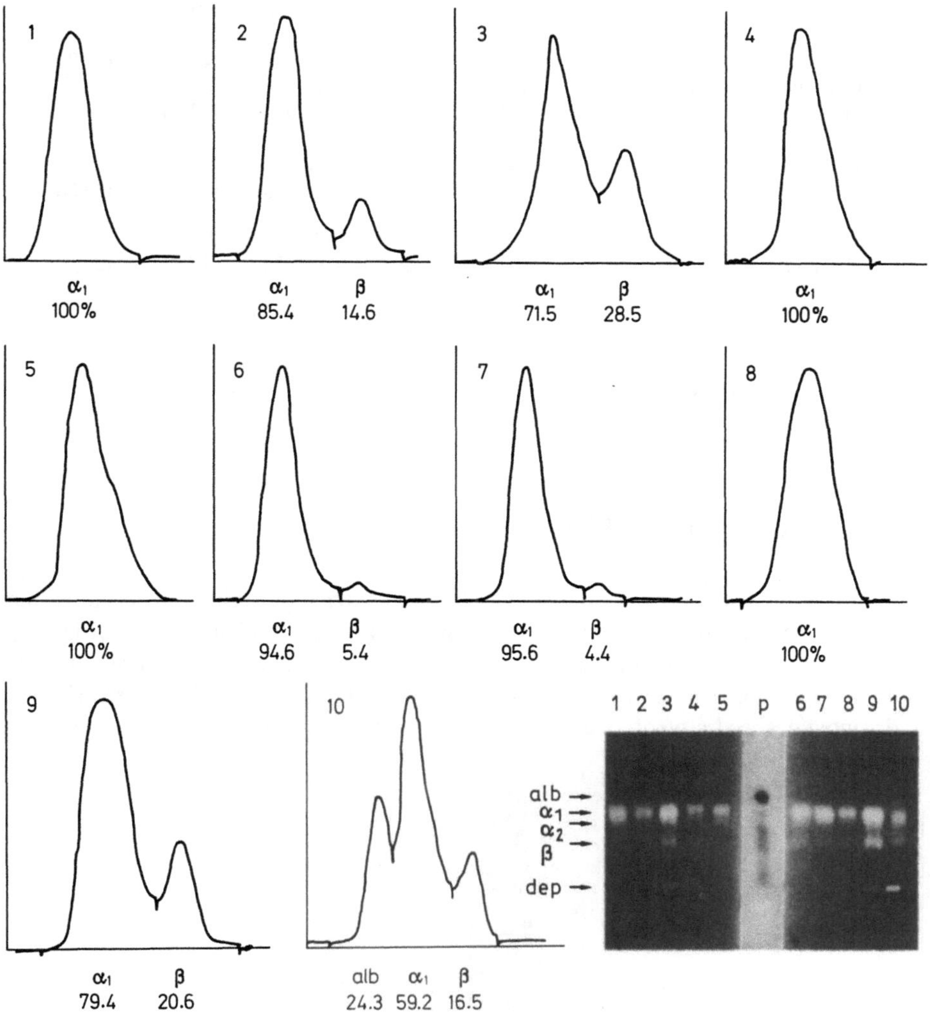

Fig. 10. GGT isoenzyme patterns in different hepatobiliary diseases. 1: chronic hepatitis; 2: cholestatic cirrhosis; 3: extrahepatic obstructive jaundice; 4 and 5: alcoholic hepatitis; 6: chronic active hepatitis; 7: cirrhosis; 8: chronic persistent hepatitis; 9: extrahepatic obstructive jaundice; 10: secondary liver tumor

Table 6. The number and frequency of the single gamma-glutamyltransferase isoenzyme bands in serum patterns from normal subjects and from subjects affected by various hepatobiliary diseases. (Reproduced with permission from: Sacchetti L, Castaldo G, Salvatore F (1988) Clin Chem 34: 353)

Diseases	N	GGT bands					
		Alb	$Alpha_1$	$Alpha_2$	Beta	Gamma	Dep
1) None	142	—	142 (100%)	142 (100%)	—	—	—
2) Chronic persistent hepatitis	13	1 (8%)	13 (100%)	1 (8%)	4 (32%)	—	—
3) Chronic active hepatitis	10	—	10 (100%)	4 (40%)	10 (100%)	2 (20%)	—
4) Cirrhosis	9	—	9 (100%)	3 (33%)	9 (100%)	2 (22%)	1 (11%)
5) Cirrhosis with intrahepatic cholestasis	14	2 (14%)	14 (100%)	—	14 (100%)	2 (14%)	2 (14%)
6) Extrahepatic obstructive jaundice	15	2 (13%)	15 (100%)	10 (67%)	15 (100%)	8 (53%)	7 (46%)
7) Liver tumor primary and secondary)	20	15 (75%)	20 (100%)	6 (30%)	20 (100%)	3 (15%)	4 (20%)
8) Non hepatic tumor	14	1 (8%)	14 (100%)	14 (100%)	—	—	—

The dash (—) indicates the absence of GGT bands

5 GGT Isoenzymes in Human Neoplasia

One of the most striking features of the GGT isoenzyme pattern [94] in serum is that most cases of hepatic carcinoma, both primary (Fig. 11) and metastatic (Fig. 12), showed an additional isoenzymatic band with a diagnostic sensitivity of about 65%. This fast-moving band usually comigrated with serum albumin. This form showed a diagnostic specificity of over 90% toward other hepatobiliary diseases and, as

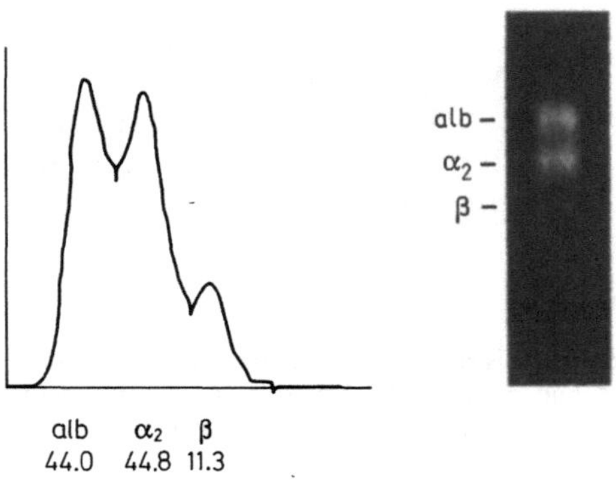

Fig. 11. GGT isoenzyme pattern of a primary liver tumor

expected, also toward neoplasias of other organs and tissues (about 92%). In fact, the pattern of the latter group coincided with that present in normal reference samples (Fig. 13). As shown in Table 6 some liver tumors contain gamma-GGT and dep-GGT. In these cases the patients showed extrahepatic obstructive jaundice caused by primary pancreatic tumors (among the cases of secondary liver tumor).

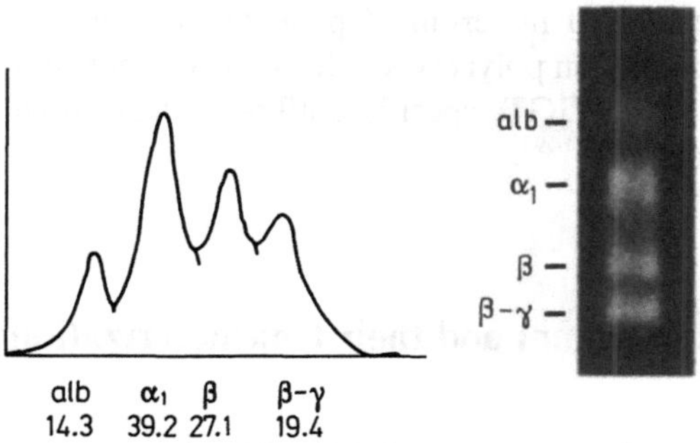

Fig. 12. GGT isoenzyme pattern of a secondary liver tumor

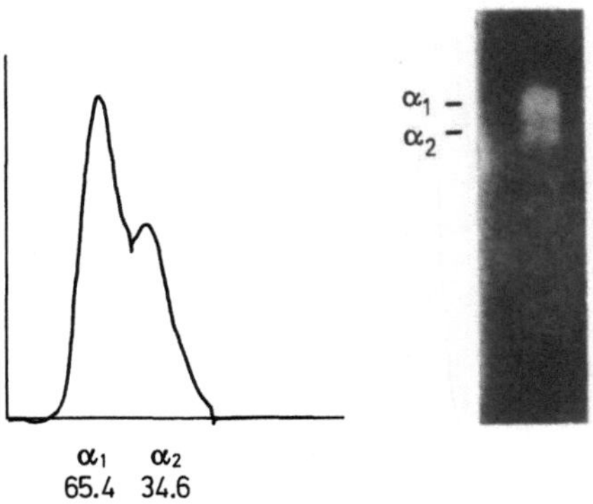

Fig. 13. GGT isoenzyme pattern of a tumor without liver involvement

In a more recent study, we examined about 50 cases of acute pancreatitis for GGT isoenzyme pattern and in most cases this practically overlapped the pattern found in extrahepatic liver jaundice (unpublished data), because the latter disorder accompanies acute pancreatitis. Rosalki [103] also indicated a beta globulin co-migrating GGT in sera from cystic fibrosis patients.

We have begun to identify the molecular structure of the alb-GGT band; unlike the $alpha_1$-GGT seen in cirrhotics, and unlike beta, gamma and dep-GGT, it does not appear as a complex between GGT and lipoproteins.

In a recent study, conducted on a smaller number of subjects, also Rosalki [95] found that the albumin comigrating GGT was associated with liver tumor, and Nemesànszky in his review [6] suggested that this isoform could be due to the action of a proteolytic enzyme on other GGT isoforms. Burlina [75], however, reported an isoenzyme with the albumin mobility in sera from patients with cirrhotic and other non-neoplastic diseases. Further studies on the molecular nature of this tumor-associated band are called for.

Carbohydrate components of GGT have been also widely studied in the last few years because different types of tumors produce forms of the enzyme with different glycosylation. Huseby and co-workers[104)] reported two GGT forms with a different degree of sialation in colorectal carcinoma tissues, and with a different affinity for Concanavalin A. Recently, Yamaguchi et al.[105)] identified a GGT with an anomalous sugar structure in carcinomatous pancreas. Isoenzymes with different glycosylations have also been identified in human hepatoma[106,107)]. Finally, some isoforms of GGT with a different glycosylation have been found in serum of patients with hepatocellular carcinoma[68–70)] using electrophoresis on polyacrylamide. With an improved ELISA method for the evaluation of serum GGT, specific antibodies have been raised against GGT from human hepatoma[108,109)].

6 The Nature of GGT Fractions in Serum and their Characterization

Our attempt to characterize the GGT fractions in connection with serum lipoproteins revealed correlations between the pattern of lipoprotein precipitation and the presence

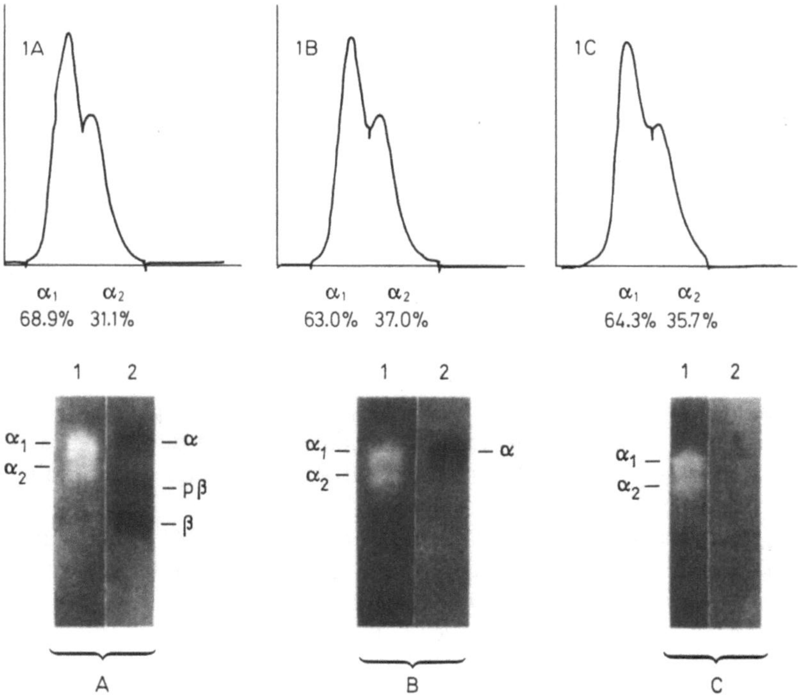

Fig. 14. GGT isoenzyme pattern (1) and lipidogram (2) in a normal subject before (A), after (B) LDL plus VLDL precipitation, and after further HDL precipitation (C). pβ: pre-beta lipoprotein. (Reproduced with permission from: Castaldo G, Fortunato G, Salvatore F, Sacchetti L (1988) It J Biochem 37: 111)

of liver neoplasia, both primary and secondary. Figure 14 shows the effect of sequential lipoprotein precipitation on the GGT pattern. The first treatment, treatment A, consisted of LDL plus VLDL precipitation, followed by electrophoresis. The results obtained on normal subjects are shown in Fig. 14, panels A, B, and C. The second treatment, treatment B, consisted of treatment A plus HDL precipitation. In all cases, after these procedures the GGT isoenzyme fractions were made visible according to the electrophoretic procedure described by Sacchetti et al. [93]. In parallel, the lipidograms from normal subjects were obtained according to the procedure described in [110]. Figure 15 shows cases of extrahepatic obstructive jaundice (1 and 1 in panel A), and the same cases after LDL plus VLDL precipitation (panel B). Figure 16 shows a cirrhotic serum where only a small part of the alpha-GGT band precipitated after treatment A, while a larger part of the alpha$_1$-GGT band was precipitated by treatment B.

All the bands precipitated with LDL plus VLDL treatment (treatment A) were associated with pre-beta and beta serum lipoproteins, whereas subsequent HDL precipitation, which produced only a decrease in the extent of the alpha$_1$-GGT band in the serum of cirrhotic patients, indicates that this alpha$_1$-GGT band is mainly associated with serum HDL.

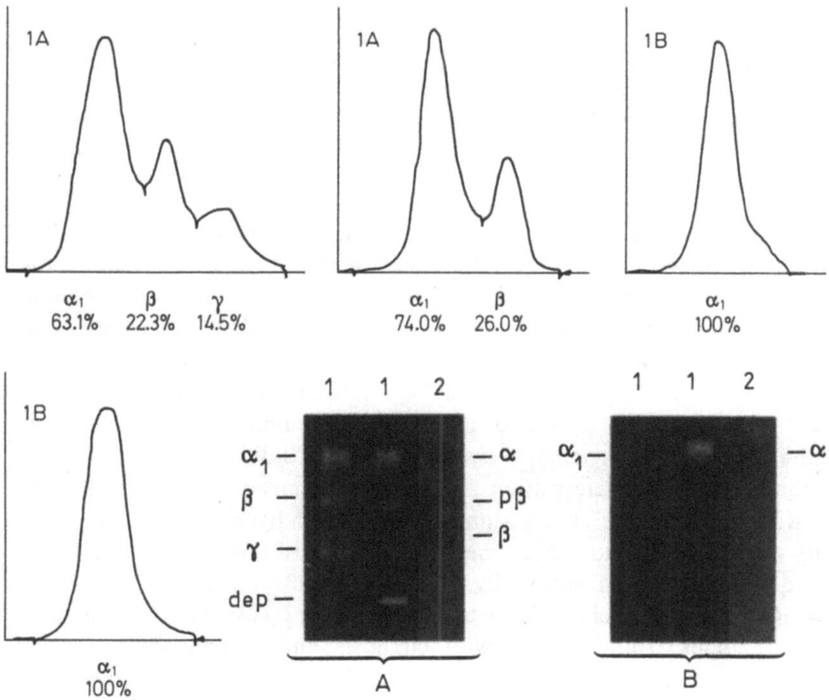

Fig. 15. GGT isoenzyme patterns in two cases of extrahepatic obstructive jaundice (1) and lipidograms from a normal subject (2), before (A) and after (B) LDL plus VLDL precipitation. pβ: pre-beta lipoprotein. (Reproduced with permission from: Castaldo G, Fortunato G, Salvatore F, Sacchetti L (1988) It J Biochem 37: 111)

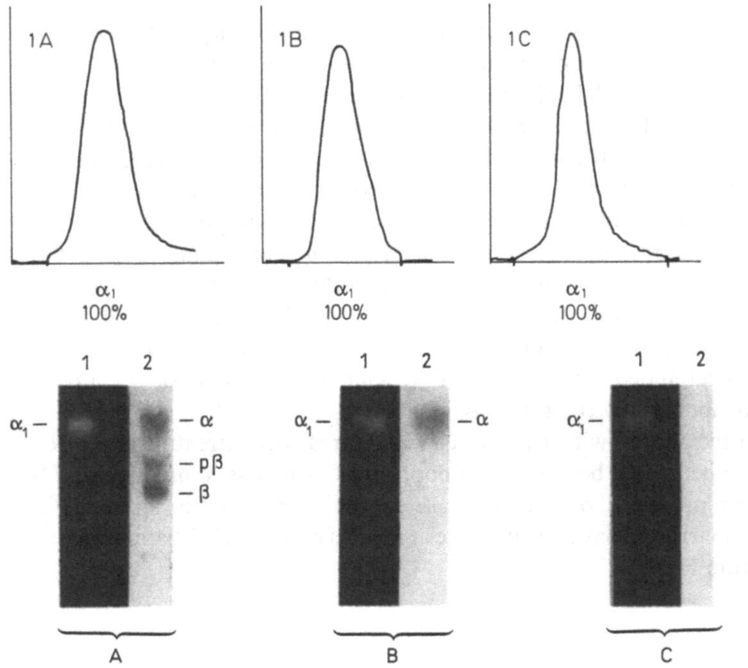

Fig. 16. GGT isoenzyme patterns in a cirrhotic (1) and lipidograms from a normal subject (2), before (A) and after (B) LDL plus VLDL precipitation, and after further HDL precipitation (C). *pβ*: pre-beta lipoproteins. (Reproduced with permission from: Castaldo G, Fortunato G, Salvatore F, Sacchetti L (1988) It J Biochem 37: 111)

Figure 17 shows serum from a neoplastic patient containing the albumin-comigrating GGT fraction. This fraction did not show any change after treatment A or after treatment B. This strongly suggests that the albumin-co-migrating fraction GGT is not associated with any major lipoprotein fraction and supports the notion that it is a lower molecular weight fraction that could derive from partial proteolytic cleavage of liver GGT as suggested by Moss [64] and Nemesànszky and Lott [6].

All these fractions are fairly stable over time. In a preliminary study [60] storage at —20 °C did not result in any change over a period of 20 days.

In addition to these characterization studies, we examined the extent of GGT precipitated with VLDL plus LDL in serum of patients with liver cancer as compared with patients with cirrhosis and chronic hepatitis. Figure 18 shows a scattergram of data obtained in this kind of study. By choosing a cut-off value of 20 U/L of precipitated GGT, it was possible to evaluate this parameter by using it to discriminate tumor-bearing patients from patients with cirrhosis and patients with chronic hepatitis. Diagnostic characteristics (see Table 7), such as sensitivity and specificity calculated according to Galen and Gambino [103], were 71% and 100%, respectively, versus normals, whereas versus patients affected by chronic hepatitis and patients with cirrhosis the values reached 87% and 80%, respectively, being 80% versus both groups. It appears, therefore, that this index, i.e. the amount of GGT complexed with LDL + VLDL [111], may contribute, particularly if further validated by other studies,

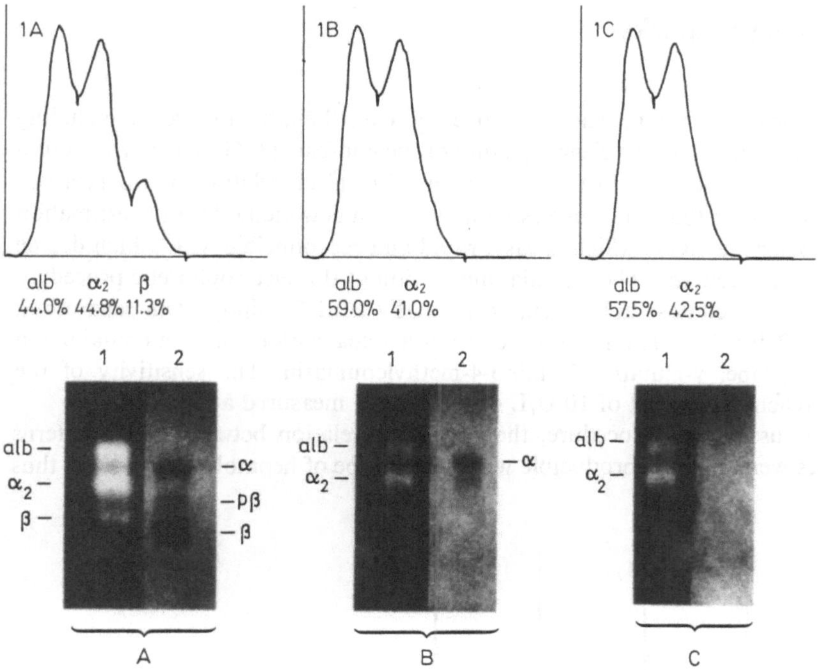

Fig. 17. GGT isoenzyme patterns in liver neoplasia (1) and lipidograms from a normal subject (2) before (A), after (B) LDL plus VLDL precipitation and after further HDL precipitation (C). (Reproduced with permission from: Castaldo G, Fortunato G, Salvatore F, Sacchetti L (1988) It J Biochem 37: 111)

to a differential diagnosis between liver cirrhosis and liver tumors versus the two diseases.

Table 7. Diagnostic sensitivity, specificity and efficiency of serum gamma-glutamyltransferase complexed with LDL + VLDL, calculated for liver malignancies toward other groups of subjects at a cut-off level of 20 U/L

Population samples			Diagnostic parameters of the tumor marker		
A		B	Sensitivity %	Specificity %	Efficiency %
Liver tumors (n = 24)	vs	normals (n = 53)	71	100	91
Liver tumors (n = 24)	vs	chronic hepatitis (n = 23)	71	87	79
Liver tumors (n = 24)	vs	cirrhosis (n = 87)	71	80	78
Liver tumors (n = 24)	vs	chronic hepatitis and cirrhosis (n = 110)	71	80	80
Liver tumors (n = 24)	vs	chronic hepatitis, cirrhosis, and normals (n = 163)	71	88	86

7 Concluding Remarks

From the review of the data that have appeared over the last three years, including data from our group, a much clearer picture of the confusing GGT isoenzyme serum pattern, starts to emerge, at least as far as its clinical correlations are concerned. The breakthrough in this field came essentially from a new method for the estimation of serum isoenzyme GGT, which is easy, rapid and reproducible with a high degree of precision and accuracy. The essential innovation of this electrophoretic procedure was the use of a cellulose acetate support usually used for lipoprotein separation, migration at 220 V for 40 mins, and the use of a visualization solution containing a fluoresecent probe, γ-glutamyl-7-amino-4-methylcoumarin. The sensitivity of the procedure reached the level of 10 U/L of total GGT measured at 25 °C.

With the use of this procedure, the clinical correlation between GGT patterns and diseases were highly reproducible within each type of hepatobiliary disease, thus

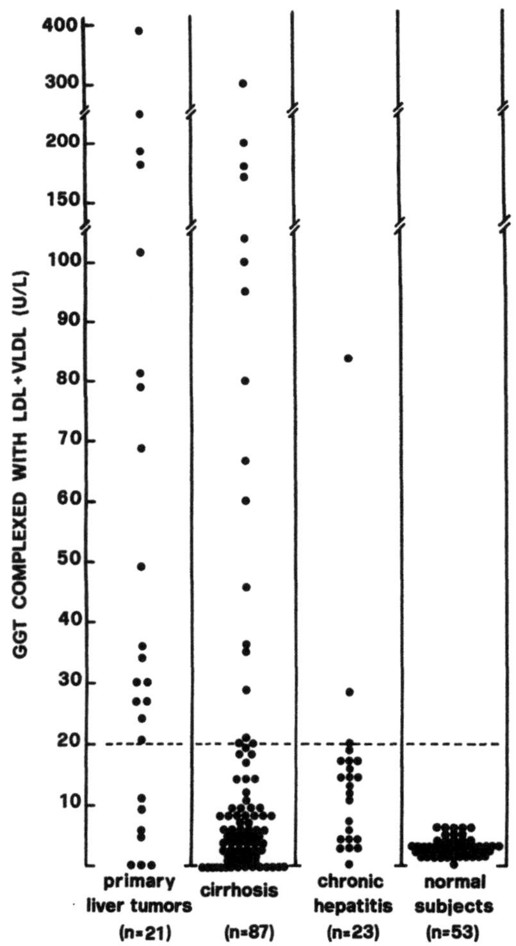

Fig. 18. GGT complexed with VLDL plus LDL, in normal subjects and in patients affected by different hepatobiliary diseases

allowing the definition of typical serum GGT isoenzyme patterns with specific liver disorders. These correlations, limited to hepatobiliary disorders, include persistent and active chronic hepatitis, cirrhosis, cholestatic cirrhosis, intra- and extrahepatic jaundice, primary and secondary liver tumor, acute pancreatitis, and non-hepatic tumors.

Data on other less specific diseases of other organs are scarce. The characterization of the various GGT isoenzyme forms was approached using sequential lipoprotein precipitation based on LDL plus VLDL, followed by HDL, precipitation. The precipitation of the GGT isoforms obtained by the first treatment (LDL plus VLDL precipitation) holds promise as a signal with which to discriminate between liver cirrhosis and tumors (Fig. 18). The discriminating potential of this approach in liver cirrhosis and tumors is probably related to differences in cholestasis between the two diseases.

All the data to emerge from this review indicate that the GGT isoenzyme pattern would be worth exploiting on a larger scale because of its correlations with hepatobiliary diseases.

8 Acknowledgements

The experimental work in the authors' laboratory was supported by grants from CNR (Rome) "Progetto Finalizzato Oncologia", the "Ministero Pubblica Istruzione" (Rome, Italy), the "Associazione Italiana per la Ricerca sul Cancro, AIRC" (Milan, Italy), and the "Regione Campania".

Thanks are due to F. Dello Stritto for his skillful technical assistance in producing the illustrations.

9 References

1. Szczeklik E, Orlowski M, Szewcsuk A (1961) Gastroenterol 41: 353
2. Goldberg DM (1979) Lab 6: 5
3. Hardison WGM (1979) in: Griffiths JC (ed) Clinical Enzymology: Gamma-glutamyltranspeptidase: A clinical overview. Masson, New York, p 199
4. Goldberg DM (1980) Crit Rev Clin Lab Sci 12: 1
5. Penn R, Worthington D (1983) Br Med J 286: 531
6. Nemesànszky E, Lott A (1985) Clin Chem 31: 797
7. Meister A, Tate SS, Ross LL (1976) in: Martonosi A (ed) The enzymes of biological membranes. Plenum Press, New York p 315
8. Tate SS, Meister A (1981) Mol Cell Biochem 39: 357
9. Meister A (1973) Science 180: 33
10. Curthoys NP, Shapiro R (1975) Febs Lett 58: 230
11. Binkley F, Wiesemann ML, Groth DP, Powell RW (1975) Febs lett 51: 168
12. Huseby NE (1978) Biochim Biophys Acta 522: 354
13. Horchi S, Inoue M, Marino Y (1978) Eur J Biochem 87: 429
14. Scherberich JE, Kleemann B, Mondorf W (1979) Clin Chim Acta 93: 35
15. Masuike M, Ogawa M, Kosaki G, Minamuira N, Yamamoto T (1982) Enzyme 27: 163

16. Lilja H, Weiber H (1985) Scand J Clin Lab Invest 43: 307
17. Yamaguchi N (1985) Clin Chim Acta 147: 75
18. Tate SS, Khadse V, Wellner D (1988) Arch Biochem Biophys 262: 397
19. Braun JP, Tainturier D, Bezille P, Raviart I, Rico AG (1984) Enzyme 31: 193
20. Sobiech KA, Wieczorek E (1981) Enzyme 26: 153
21. Tate SS, Ross ME (1977) J Biol Chem 252: 6042
22. Huseby NE (1977) Biochim Biophys Acta 483: 46
23. Shaw LM, London JW, Petersen LE (1978) Clin Chem 24: 905
24. Shaw LM, Petersen Archer L, London JW, Marsh E (1980) Clin Chem 26: 1523
25. Sundarsingh Daniel D, Ramakrishna BS, Balasubramanian KA (1987) Clin Chim Acta 162: 319
26. Goodman SI, Mace JW, Pollak S (1971) Lancet i: 234
27. Schulman JD, Goodman SI, Mace JW, Patrick AD, Tietze F, Butler E (1975) Biochem Biophys Res Comm 65: 68
28. Rambabu K, Pattabiraman TN (1976) Clin Chim Acta 73: 251
29. Rambabu K, Jacob RT, Pattabiraman TN (1981) Clin Chim Acta 109: 257
30. Salgo L, Szabo A (1982) Clin Chim Acta 126: 9
31. Linder M, Sudaka P (1982) Clin Chim Acta 118: 77
32. Jung K, Wischke UW (1984) Clin Chem 30: 856
33. Jung K, Pergande M, Reinholdt C, Schulze B, Scholz D, Strobelt V (1984) J Clin Chem Clin Biochem 22: 523
34. Amodio P, Bazzerla G, Malatesta R, Gatta A (1985) Enzyme 33: 216
35. Ruiz-Moral R, Perez-Jimenez F, Cosano A, Hens M, Munoz L, Jimenez Alonso J, Munoz R, Pereperez JJ (1985) Enzyme 33: 105
36. Jimenez-Alonso J, Jaimez L, Perez-Jimenez F, Grilo A, Jimenez-Pereperez JA (1982) Clin Chem 28: 1988
37. Wenham PR, Price CP, Sammons HG (1978) J Clin Pathol 31: 666
38. Indirani N, Hill PG (1977) Biochim Biophys Acta 483: 57
39. Wenham PR, Horn DB, Smith AF (1981) Clin Chim Acta 112: 113
40. Bulle F, Mattei MG, Pawlak A, Passage E, Chobert MN, Laperche Y, Guellaen G (1987) Human Genet 76: 283
41. Laperche Y, Bulle F, Aissani T, Chobert MN, Aggerbeck M, Hanoune J, Guellaen G (1986) Proc. Natl Acad Sci 83: 937
42. Coloma J, Pitot HC (1986) Nucl Acids Res 14: 1393
43. Tate SS, Nash B (1987) Febs Lett 211: 133
44. Tate SS, Meister A (1977) Proc Natl Acad Sci (USA) 74: 931
45. Hornichi S, Inoue M, Morino Y (1978) Biochem Biophys Res Comm 80: 873
46. Tate SS (1986) Febs Lett 194: 33
47. Nash B, Tate SS (1982) J Biol Chem 257–585
48. Nash B, Tate SS (1984) J Biol Chem 259–678
49. Matsuda Y, Tsuji A, Kuno T, Katunuma N (1983) J Biochem 94: 755
50. Altman RA, Hughey RP (1986) Biochem Intern 13: 1009
51. Matsuda Y, Tsuji A, Katunuma N (1983) J Biochem 93: 1427
52. Miura T, Matsuda Y, Tsuji A, Katunuma N (1981) J Biochem 89: 217
53. Tsuchida S, Imai F, Sato K (1981) J Biochem 89: 775
54. Das ND, Shichi H (1979) Life Sci 25: 1821
55. Huseby NE (1982) Clin Chim Acta 124: 103
56. Wenham PR, Horn DB, Smith AF (1985) Clin Chem 31: 569
57. Huseby NE (1986) Adv Clin Enzymol 3: 67
58. Wenham PR (1986) Adv Clin Enzymol 3: 74
59. Artur Y, Sanderink GJ, Maire J (1987) Ann Biol Clin 45: 277
60. Castaldo G, Fortunato G, Salvatore F, Sacchetti L (1988) It J Biochem 37: 111
61. Artur Y, Wellman-Bednawska M, Jacquier A, Siest G (1984) Clin Chem 30: 631
62. Artur Y, Wellman-Bednawska M, Jacquier A, Siest G (1984) Clin Chem 30: 1318
63. De Broe ME, Borgers M, Wieme RJ (1975) Clin Chem Acta 59: 369
64. Moss DW (1984) in: Goldberg DM, Werner M (eds) Selected Topics in Clin Enzymol 2: 517

65. Moss DW, Echetebu ZO, Whitaker KB, Traynor OJ, Wood CB (1982) in: Siest G, Heusghem C (eds) Gamma-glutamyltransferases. Adv Biochem Pharmacol 3rd series, p 41
66. Echetebu ZO, Moss DW (1982) Enzyme 27: 1
67. Echetebu ZO, Moss DW (1982) Enzyme 27: 9
68. Ishii H, Yasuraoka S, Shigeta Y, Takagi S, Kamiya T, Okuno F, Miyamoto K, Tsuchiya M (1978) Life Sci 23: 1393
69. Horner F, Kellen JA, Kingstone E, Maharaj N, Malkin A (1979) Enzyme 24: 217
70. Kok PJMJ, Seidel B, Holtkamp HC, Huisman J (1978) Clin Chim Acta 90: 209
71. Braide SA, Davies TJ (1987) Ann Clin Biochem 24: 391
72. Artur Y, Wellman-Bednawska M, Siest G (1984) Ann Biol Clin 42: 115
73. Bornschein W (1975) Clin Chim Acta 61: 325
74. Laursen T, Jacyszyn K (1968) Clin Chim Acta 21: 497
75. Burlina A (1978) Clin Chem 24: 502
76. Rosalki SB, Nemesànszky E, Foo AY (1981) Ann Clin Biochem 18: 25
77. Park HR, Geitner AM, Fritsche HA (1984) Clin Chem 30: 1010
78. Hammond KD, Gravenor MJ, Piesas D (1985) Ann Clin Biochem 22: 376
79. Burlina A, Secchiero S, Bertorelle R, Zaninotto M (1987) Med Sci Res 15: 499
80. Hetland O, Andersson TR, Gerner T (1975) Clin Chim Acta 62: 425
81. Echetebu ZO, Moss DW (1979) Clin Chim Acta 95: 433
82. Wenham PR, Price CP, Sammons HG (1978) Ann Clin Biochem 15: 146
83. Koima J, Kanatani M, Nakamura N, Kashiwagi T, Tohjoh F, Akiyama M (1980) Clin Chim Acta 106: 165
84. Sawabu N, Nakagen M, Ozaki K, Wbkabayashi T, Toya D, Hattori N, Ishii M (1983) Cancer 51: 327
85. Kew MC, Wolf P, Whittaker D, Rowe P (1984) Br J Cancer 50: 451
86. Sinha PK, Righetti PG (1986) J Biochem Biophys Meth 12: 289
87. Sinha P, Kottgen E, Righetti PG (1988) Fres Z Anal Chem 8216: 1
88. Lilja H, Jeppsson JO, Kristensson H (1983) Clin Chem 29: 1034
89. Delanghe J, Buyzere M, Faust U, Krause-Jarres J, De Scheerder I (1987) Enzyme 83: 22
90. Delanghe JR, Buyzere ML, De Scheerder IK, Claeys LR, Wieme RJ (1987) Clin Chim Acta 162: 311
91. Tsuji A, Matsuda Y, Katunuma N (1980) Clin Chim Acta 104: 361
92. Hoskins JA, Davies FF (1986) Anal Biochem 152: 314
93. Sacchetti L, Castaldo G, Fortunato G, Salvatore F (1988) Clin Chem 34: 419
94. Sacchetti L, Castaldo G, Salvatore F (1988) Clin Chem 34: 352
95. Rosalki SB (1982) in: Siest G, Heusghem C (eds) Gamma-glutamyltransferases. Adv Biochem Pharmacol 3rd series, p 147
96. Brocklenurst D, Wilde CE (1980) Clin Chem 26: 588
97. Buamah PK, Skillen AW, Davison V (1984) Clin Chim Acta 141: 285
98. Carbarns NJ, Gasden C, Brock DJH (1983) Lancet i: 329
99. Jalanko H, Ranta T, Lehtonem E, Roaslahti E (1983) Clin Chim Acta 134: 337
100. Moniz C, Nicolaides KH, Keys D, Rodeck CH (1984) J Clin Pathol 37: 700
101. Garcia MP, Tutor JC, Sanjose M, Porto JA, Fraga JM, Paz JM, Rodriguez-Segade S (1987) Clin Biochem 20: 269
102. Galen RS, Gambino SR (1976) in: Galen SR, Gambino SR (eds) Beyond normality: the predictive value and efficiency of medical diagnoses, Wiley and Sons, New York
103. Rosalki SB, Foo AY, Hjelm M, Dinwiddie R (1985) Clin Chem 31: 779
104. Huseby NE, Eide TJ (1983) Clin Chim Acta 135: 301
105. Yamaguchi N, Kawai K, Ashihara T (1986) Clin Chim Acta 154: 133
106. Yamamoto H, Sumikawa K, Hada T, Higashino K, Yamamura Y (1981) Clin Chim Acta 111: 229
107. Tsuchida S, Yamazaki T, Cambo EM, Morita T, Matsue H, Yoshida Y, Sato K (1985) Clin Chim Acta 152: 17
108. Taniguchi N, House S, Kuzumaki N, Yokosawa N, Yamagiwa S, Lizuka S, Makita A, Sekiya C (1985) J Natl Cancer Inst 75: 841
109. Taniguchi N, Lizuka S, Nai Zhe Z, Honse S, Yokosawa N, Ono M, Kinoshita K, Makita A, Sekiya C (1985) Cancer Res 45: 5835

110. Chin HP, Scholnick HR, Morfin R (1970) J Lipid Res 11: 583
111. Sacchetti L, Castaldo G, Cimino L, Budillon G, Salvatore F (1988) Clin Chim Acta 177: 167
112. Pawlak A, Lahuna O, Bulle F, Suzuki A, Ferry N, Siegrist S, Chikhi N, Chobert MN, Guellaen G, Laperche Y (1988) J Biol Chem 263: 9913
113. Rajpert-De Meyts E, Heisterkamp N, Groffen J (1988) Proc Natl Acad Sci USA 85: 8840

Alkaline Phosphatase in Hepatobiliary Disease

Donald W. Moss

Royal Postgraduate Medical School, Hammersmith Hospital, London W12 ONN, U.K.

The elevated activity of alkaline phosphatase in serum in hepatobiliary disease has been of established diagnostic value for nearly 60 years. The contrast between the marked elevations typically associated with obstructive jaundice and the slight to moderate elevations in parenchymatous disease such as hepatitis remains a useful generalization in the interpretation of alkaline phosphatase levels in liver disease. However, a satisfactory explanation for the changes seen in alkaline phosphatase levels in hepatobiliary disease and for the absence of a close correlation with serum bilirubin took 30 years to emerge. It is now recognized that interference with bile-flow stimulates new synthesis of alkaline phosphatase in hepatocytes, although the exact molecular mechanism of enzyme induction and release remains to be elucidated.

The structural gene that encodes liver alkaline phosphatase is widely expressed, e.g. also in bone and kidney. Tissue-specific variations in glycosylation confer individual properties on the alkaline phosphatases from these tissues which allow them to be identified and measured. Modifications during or after enzyme release may give rise to further isoforms potentially useful in diagnosis.

The liver also clears adult-intestinal alkaline phosphatase from the plasma at what is normally a rapid rate, mediated by uptake by galactosyl-glycoprotein receptors on the hepatocyte surface. Therefore, in contrast to hepatic phosphatase, the level of intestinal phosphatase in plasma is controlled by its rate of clearance from the circulation and not by its rate of entry into it. Liver disease may reduce the rate of clearance of intestinal phosphatase. Intestinal alkaline phosphatase reaches the plasma by way of the lymph. Comparison of intestinal phosphatase in the plasma with the isoenzyme extracted from fresh intestinal tissue shows the former has lost its membrane-binding domain, presumably by proteolytic action in the intestine. In contrast, hepatic phosphatase retains its membrane-binding domain and can form complexes with lipids in bile and plasma.

Progress in Clinical Biochemistry
and Medicine, Vol. 8

1 Introduction

The measurement of serum alkaline phosphatase activity entered the list of biochemical liver function tests as the result of a chance observation by W. M. Roberts in 1930. The inclusion of two patients with obstructive jaundice and one with catarrhal jaundice (infective hepatitis) in a series of patients with bone disease in whom serum alkaline phosphatase was being measured revealed a marked elevation in obstructive jaundice and only a minimal elevation in infective hepatitis [1]. Thus, the association between a raised serum alkaline phosphatase and impaired biliary excretion, and particularly the more marked elevation in extrahepatic obstruction which still remains a useful generalization in the interpretation of this test, was established on the basis of just three patients. Although the difference in alkaline phosphatase levels between hepatogenic and post-hepatic jaundice was soon confirmed, the absence of a close correlation between serum bilirubin and alkaline phosphatase and, in particular, the rise in alkaline phosphatase in the absence of jaundice, seen for example in patients with malignant disease affecting the liver, presented an apparent paradox (Fig. 1). Thus, in contrast to the rapidity with which Roberts' original observation was confirmed and established in diagnosis, a satisfactory explanation of it took 30 years to emerge.

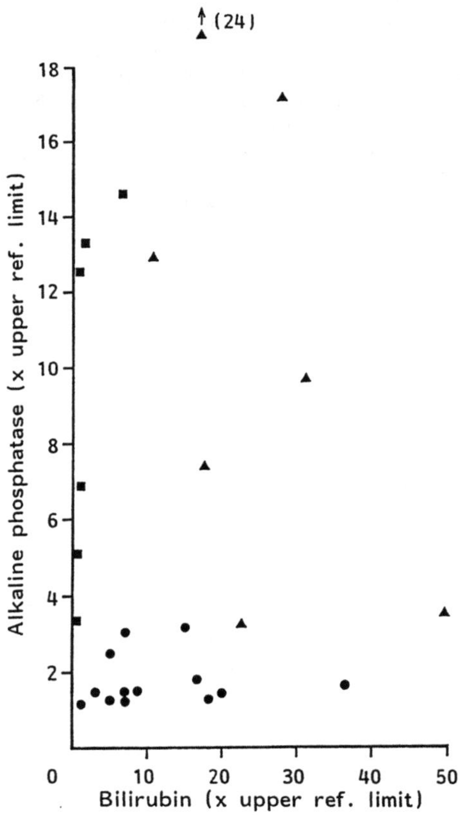

Fig. 1. Typical serum alkaline phosphatase activities and bilirubin concentrations in hepatobiliary disease, both expressed as multiples of the upper reference limits. The contrast between alkaline phosphatase levels in infective hepatitis (●) and in obstructive jaundice (▲) is seen, but the dissociation between alkaline phosphatase and bilirubin, particularly in secondary liver cancer (■) is also apparent

The long debate between the retention theory — i.e., the supposed failure to excrete alkaline phosphate derived mainly from bone — and the theory of the hepatic origin of the increased plasma alkaline phosphatase was finally resolved in favour of the latter view by two lines of evidence. Experiments in whole animals with obstructions of bile-ducts draining all or part of the liver [2,3,4], or in isolated perfused liver [5], showed that the liver responds to biliary obstruction by new synthesis of alkaline

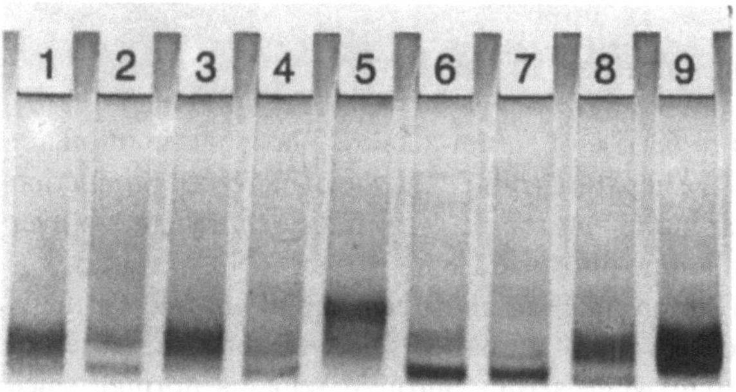

Fig. 2. Alkaline phosphatase zones in sera separated by electrophoresis in polyacrylamide gel at pH 9.5. The compact, most anodal zone of liver phosphatase is the most intense zone in samples 6 and 7. The less anodal, more diffuse bone phosphatase zone is prominent in samples 1, 3, 8 and 9, and bone and liver zones are both present in varying proportions in samples 1, 2, 4, 6 and 8. The more cathodal intestinal phosphatase zone is seen in sample 5, in which it occurs together with the bone phosphatase zone. The anodal direction is downwards. (From Moss DW, Electrophoresis of human alkaline and acid phosphatases. In: Wolf PL, (ed) Electrophoresis of serum proteins and isoenzymes, Clinics in Laboratory Medicine, Vol 6, p 507, WB Saunders, Philadelphia 1986)

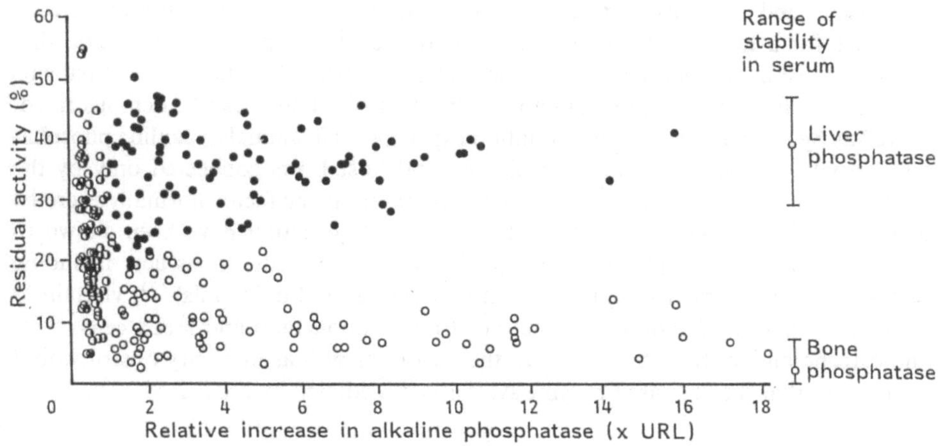

Fig. 3. Heat-stability of alkaline phosphatase (expressed as percentage residual activity after 10 min incubation at 56 °C) as a function of total serum alkaline phosphatase activity (as multiples of upper reference limit, URL) in sera of normal individuals (○) and patients with osteoblastic bone disease (○) or hepatobiliary disease (●). The low and high residual activities in bone and hepatobiliary diseases reflect, respectively, the increasing amounts of the less-stable bone alkaline phosphatase and more-stable liver alkaline phosphatase in the serum as the total activity rises. The ranges (±2 S.D.) of residual activities expected in sera containing exclusively either bone or liver alkaline phosphatase are shown at right

phosphatase. The second category of evidence favouring the hepatogenic theory was the demonstration that tissue-specific properties of liver alkaline phosphatase, such as electrophoretic mobility [6] and heat stability [7], are exhibited by the alkaline phosphatase present in increased amounts in serum in hepatobiliary disease (Figs. 2 and 3), not those of bone phosphatase as predicted by the retention theory.

2 Induction of Alkaline Phosphatase in the Liver

The main causative factor in the rapid rise of serum alkaline phosphatase in biliary obstruction is the new synthesis of the hepatic enzyme. The site of increased production of the enzyme is the hepatocytes [8], and the recent work indicates that in rat liver there is increased translation of alkaline-phosphatase messenger RNA [9]. The discovery of the relationship between serum alkaline phosphatase and enzyme synthesis explains the paradoxical clinical observation that patients may exhibit a raised serum alkaline phosphatase of liver origin while serum bilirubin remains normal, since localized biliary obstruction, e.g. by a tumour, will stimulate alkaline phosphatase production locally while bilirubin excretion continues essentially unimpaired. Enzyme induction also removes the main objection to the hepatogenic theory; i.e. that the levels of alkaline phosphatase in normal liver are sufficient to account for the elevations seen in plasma in obstructive liver disease.

The nature of the inducer of alkaline phosphatase in the hepatocytes remains unknown, although the observation that bile acids can induce alkaline phosphatase in rat hepatocytes in culture [10] suggests that the accumulation of these compounds in plasma and liver as a result of obstruction may be the causative factor. Evidence for a humoral factor is provided by the often-repeated observation that alkaline phosphatase of apparent hepatic origin is increased in the plasma of patients with renal-cell carcinoma in whom no involvement of the liver can be demonstrated, as well as by auxiliary liver-transplantation experiments in animals: alkaline phosphatase activity increased in both transplanted and host livers connected only by the bloodstream, suggesting the presence of a humoral inducing factor normally excreted in the bile [11]. Various stimuli other than biliary obstruction have been shown to stimulate alkaline phosphatase activity in non-hepatic cells or tissues, including dietary phosphate (rat kidney) [12], vitamin A (mouse-tail epidermis) [13], vitamin D (chick intestinal brush-border) [14], steroids (mouse duodenum and HeLa cells) [15,16] and catecholamines (rat atrium) [17]. In these cases as well as in biliary obstruction it is probable that the increased enzyme synthesis is mediated by cyclic AMP [18].

3 Release of Alkaline Phosphatase from Hepatocytes

Although induction of alkaline phosphatase in hepatocytes is the main determinant of the raised serum alkaline phosphatase of hepatobiliary disease, this normally membrane-bound enzyme must nevertheless also be released in increased amounts from the cells before it manifests itself as a raised activity in the circulation.

That increased release of membrane-bound enzymes can proceed in the absence of immediate new enzyme synthesis can be seen in the case of γ-glutamyl transferase and 5′-nucleotidase in bile duct-ligated animals: activities of these enzymes increase in serum, without the immediate rise in levels within the tissue that is seen in the case of alkaline phosphatase [4]. Conversely, increased alkaline phosphatase in the liver but not in serum has been demonstrated in partially-hepatectomized and carbon tetrachloride-poisoned rats [19]. Both increased enzyme synthesis and increased enzyme release therefore can be presumed to contribute to the changes in serum alkaline phosphatase in hepatobiliary disease.

Alkaline phosphatase is recovered mainly in the plasma-membrane fraction of normal rat hepatocytes following separation of the liver into different cell types and isopycnic centrifugation [20]. After synthesis of the enzyme protein, it undergoes post-translational modification such as the addition of carbohydrate side-chains before incorporation into the plasma membrane. The enzyme is normally localized mainly at the canalicular surface of the cells in the rat [21], probably partly or mainly as an ectoenzyme. From this location it can be shed into the bile, either eluted by the detergent action of the bile or as a result of membrane recycling, forming high

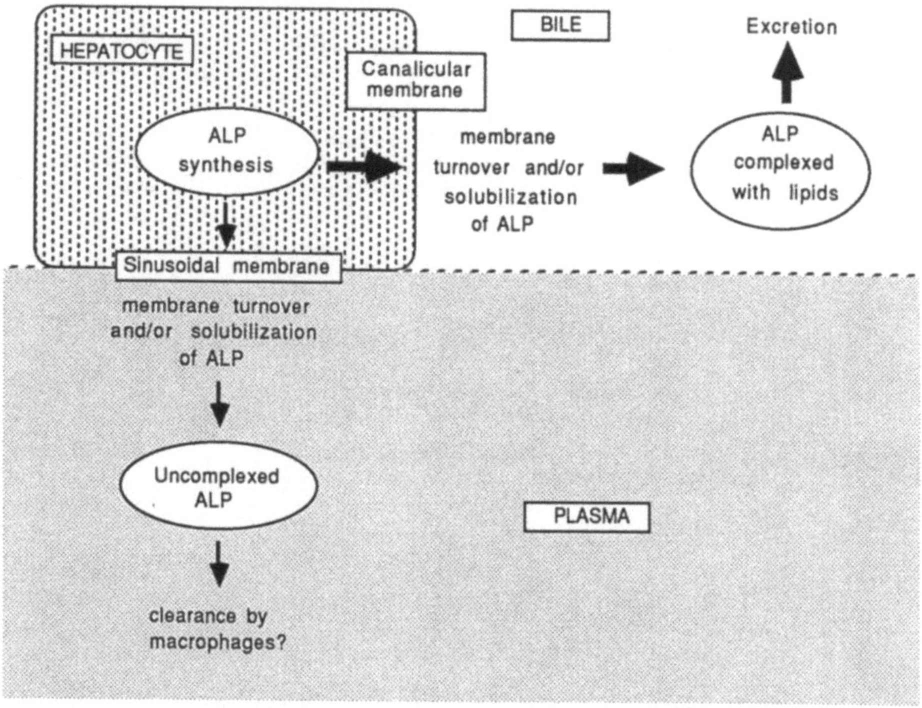

Fig. 4. Probable normal metabolic pathways of hepatic alkaline phosphatase (ALP). After synthesis in the hepatocytes most of the enzyme passes to the canalicular membrane and thence into the bile, either as a result of membrane turnover or the solubilizing action of bile salts. Non-covalent complexes with lipids are formed during or after this process, and the complexes are excreted with the bile. Some alkaline phosphatase passes to the sinusoidal membrane and into the plasma in an uncomplexed form and is cleared by tissue macrophages

molecular-mass complexes with lipid (Fig. 4). The alkaline phosphatase released accounts for almost all the phosphatase activity of the bile and, in the absence of cholestasis, passes into the intestine. Some alkaline phosphatase is also released directly into the plasma from sinusoidal membranes (Fig. 4).

Increasing extra- or intracellular bile acid concentrations as a result of biliary obstruction may solubilize from membranes the increased amounts of alkaline phosphatase being produced, or prevent incorporation into binding sites in the plasma membrane. Alkaline phosphatase entering the bile can, in biliary obstruction, find its way back into the circulation, most probably by regurgitation through the tight junctions between adjacent hepatocytes, which become more permeable in cholestasis (Fig. 5). However, direct entry of alkaline phosphatase into the plasma from the hepatocytes is also increased and is quantitatively the more important route (Fig. 5). There is evidence that the distribution of alkaline phosphatase in rat hepatocytes changes as a result of cholestasis, from localization on the exterior surfaces of the bile-canalicular membrane to a generalized membrane- and intracellular distribution [22], from which it is more likely to pass directly to the circulation. In human

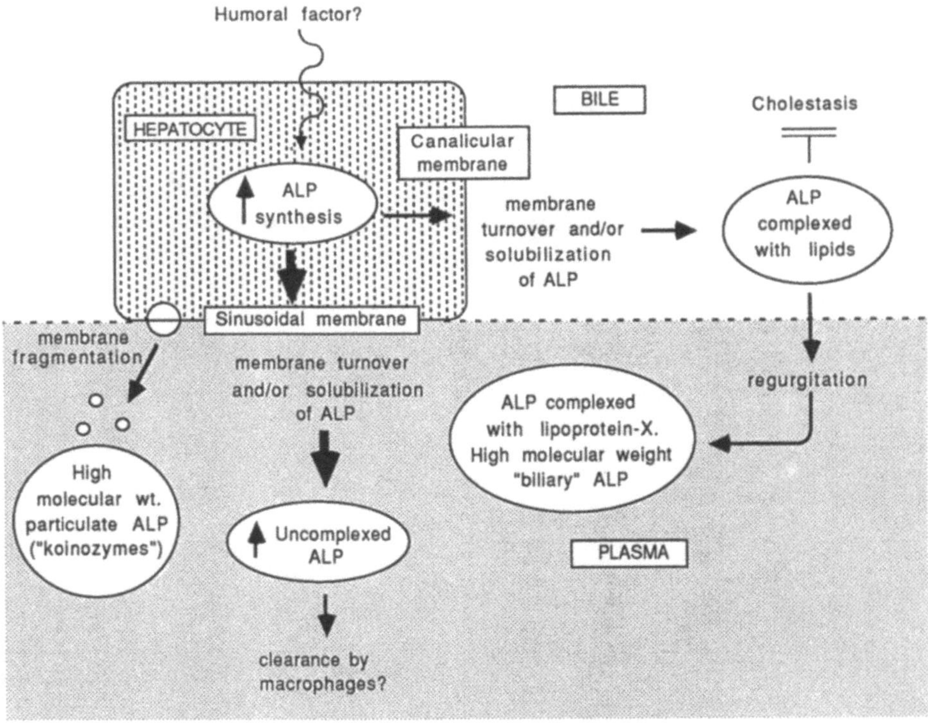

Fig. 5. Probable changes in metabolism of hepatic alkaline phosphatase (ALP) in cholestasis. Alkaline phosphatase synthesis in hepatocytes is increased, and as a result of redistribution from canalicular to sinusoidal membranes, more enzyme passes directly to plasma, where the activity of the uncomplexed enzyme is correspondingly increased. Complexes of alkaline phosphatase formed in the bile are regurgitated into plasma where they can be detected as high molecular-mass complexes with lipoprotein-X. Additional high molecular-mass forms of alkaline phosphatase consisting of membrane fragments ("koinozymes") may also occur as a result of hepatocyte damage or destruction

liver, also, extrahepatic biliary obstruction is characterized by intense enzyme activity along sinusoids [23].

4 Molecular Forms of Hepatic Alkaline Phosphatase in Serum

The possible identification of organ-specific enzyme forms in plasma first attracted the attention of clinical enzymologists to the study of isoenzymes. The demonstration that, in hepatobiliary disease, plasma alkaline phosphatase exhibits the same properties as the enzyme extracted from liver tissue was an important confirmation of the validity of the hepatogenic theory, as already noted. The tissue-specific properties of hepatic alkaline phosphatase are the result of differential glycosylation of the product of a structural gene that is also expressed in bone, kidney and other tissues [24]: thus, they reflect the operation of normal processes of differentiation and function. However, post-translational modifications of enzyme molecules occurring during or after their release from cells may provide clues to the nature of pathological processes. The altered pattern of metabolism of hepatic alkaline phosphatase suggested in Fig. 5 predicts that various forms of the enzyme will be found in plasma or serum in hepatobiliary disease.

4.1 Uncomplexed Hepatic Alkaline Phosphatase

The main component of alkaline phosphatase in serum in hepatobiliary disease of all types has the same properties of electrophoretic mobility in both sieving [6] and non-sieving [25] media, and the same kinetic properties [26], molecular size determined by gradient-pore electrophoresis [27] and stability to heat [7] as alkaline phosphatase extracted from human liver, e.g. by treatment of the tissue with aqueous *n*-butanol. This similarity between the serum and tissue-derived enzymes is perhaps the most convincing evidence that alkaline phosphatase can pass directly from cell to plasma: an obligatory passage into the bile followed by regurgitation into the plasma would be expected to modify the alkaline phosphatase through processes such as aggregation. (As mentioned later, forms of alkaline phosphatase making up a smaller proportion of the total activity are also present in serum in liver disease which probably do correspond to regurgitated enzyme).

The alkaline phosphatase extracted from liver tissue with *n*-butanol is sensitive to digestion with bromelain, which reduces its apparent molecular mass by approximately 8 kDa, from 189 to 181 kDa, without a significant reduction in phosphatase activity [27]. A form with a molecular mass of 181 kDA can also be obtained by treating liver tissue with bromelain (Fig. 6). The fragment of 8 kDa removed by bromelain thus appears to correspond to part of the hydrophobic domain that is presumed to anchor the alkaline phosphatase molecule to the cell membrane. The form of alkaline phosphatase that predominates in serum in hepatobiliary disease corresponds in size to that extracted from liver tissue with *n*-butanol, and is also sensitive to digestion with bromelain [27], undergoing a reduction in molecular mass similar to that of the enzyme from liver tissue on treatment (Fig. 6). This implies that hepatic alkaline phosphatase is released from hepatocytes complete with all or part of its membrane-binding domain, favouring a mechanism which involves solubilization from membranes by detergent action, and/or failure of incorporation into membranes, rather than alternative possibilities such as proteolytic cleavage. Alkaline phosphatase is bound to cell mambranes through a carboxy-terminal glycosyl-phosphatidylinositol (GPI) structure [58]. It is not yet known how much of the CPI "tail" is present in alkaline phosphatases solubilized in various ways, nor where the bromelain-sensitive bond is located.

Identification of uncomplexed hepatic alkaline phosphatase, which usually appears as the most prominent alkaline phosphatase zone after electrophoresis of sera of patients with hepatobiliary disease, by the inspection of zymograms provides the answer to about half of all requests for alkaline phosphatase isoenzyme analysis; such as the origin of an unexpected elevation in serum — a not uncommon finding if biochemical profiling is practised. However, quantitative methods are needed when looking for minimal changes in one tissue-specific enzyme form, especially when the form of interest is a quantitatively minor component, or when following the time-course of changes. Quantitation of bone and liver phosphatases can be based on electrophoresis under conditions of improved resolution, such as after brief neuraminidase-treatment [28] or in the presence of wheat-germ lectin [29], or on selective precipitation of liver phosphatase with acidic ethanol [30] or of bone phosphatase with wheat-germ lectin [29].

The difference in heat-stability between bone and liver phosphatases has been made the basis of their separate quantitation in serum. The absolute rates of inactivation of bone and liver phosphatases in individual sera are affected by various sample-dependent factors, principally pH, although the relative difference between the two enzyme forms remains [31,32]. These sample-dependent effects can be eliminated by

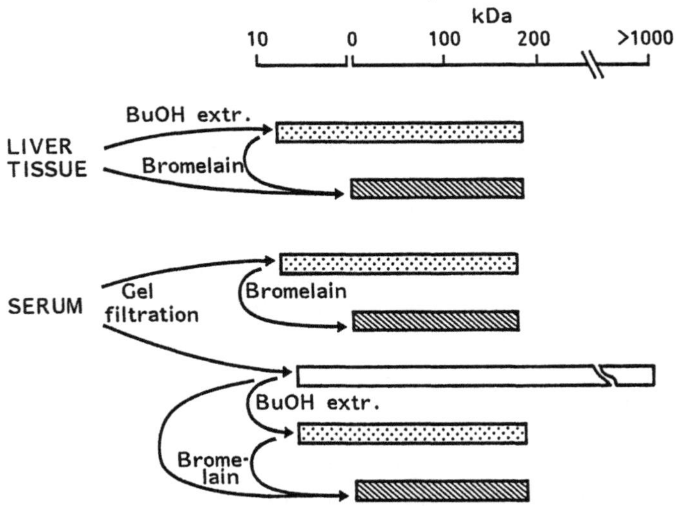

Fig. 6. Comparative molecular sizes determined by gradient-pore electrophoresis of various forms of human hepatic alkaline phosphatase showing the effect of digestion with bromelain. Bromelain removes a fragment of approximately 8 kDa from the tissue enzyme, which is presumed to correspond to part of a hydrophobic, membrane-binding domain. This domain is also present in the lower-molecular-mass fraction recovered from serum by gel-filtration. High-molecular-mass alkaline phosphatase from serum also contains the bromelain-cleavable, membrane-binding domain. (Note difference in scale to left of bromelain-cleavage point, 0)

making several measurements of residual activity during heat-inactivation, e.g. at 56 °C, so that the half-inactivation time of liver phosphatase can be determined for each sample. A minimum of three measurements, at 0, 15, and 25 min, has been selected by experiment [33]. This procedure has been extensively evaluated and shown to be robust and reliable, given careful technique, and can be regarded as a reference method. The between-run coefficient of variation is 6% for the determination of liver phosphatase at a level of 2–3 times the adult upper reference limit [34].

4.2 High-molecular-mass Alkaline Phosphatase

Plasma or serum from normal individuals contains a trace (less than 5% of total activity) of alkaline phosphatase of high apparent molecular mass, i. e. it is excluded from Sepharose 6B during gel-filtration [35,36]. The amount and proportion of this fraction is frequently raised in hepatobiliary diseases. This high-molecular-mass fraction also carriers a high net negative charge and therefore migrates rapidly towards the anode on non-sieving electrophoresis. The latter property has given rise to the name "fast-liver" fraction. However, since high molecular mass is an equally distinctive property and one which is more obviously related to its origins, the term "high-molecular-mass fraction" is preferred here. Although some properties of high-molecular-mass alkaline phosphatase, such as heat-stability, may be slightly altered compared with those of the lower-molecular-mass forms [37], these differences are not sufficiently great to suggest that the high-molecular-mass form is other than a modified form of hepatocyte alkaline phosphatase. When the high-molecular-mass fraction is recovered from serum by gel-filtration and extracted with *n*-butanol its molecular mass corresponds to that of alkaline phosphatase extracted in a similar way from liver tissue (Fig. 6). Furthermore, treatment with bromelain of either the high-molecular-mass fraction or the enzyme extracted from it produces an alkaline phosphatase resembling in size the bromelain-solubilized enzyme from liver [27] (Fig. 6).

The rapidly-migrating, high-molecular-mass alkaline phosphatase is usually not apparent on electrophoresis of normal sera. However, this fraction is readily detectable on electrophoresis, or is found to be increased on gel-filtration, in as much as 90% of sera from patients with cholestasis [38,39]. Since total alkaline phosphatase is inevitably also raised in these patients, the increase in diagnostic information obtained by the demonstration of increased high-molecular-mass alkaline phosphatase is perhaps limited. However, a particularly important role for high-molecular-mass alkaline phosphatase has been suggested in the detection of hepatic metastases [35,40]. The incidence of an increased high-molecular-mass fraction is indeed high in sera from patients with hepatic metastases, although the proportion of apparently false positives — raised values in patients in whom hepatic metastases are not otherwise detectable — seems also to be high. For example, in a series of patients with colorectal cancer, high-molecular-mass alkaline phosphatase was raised in sera from 14 out of 16 in whom metastases were detected by bimanual palpation of the liver, but was also raised in 13 out of 26 patients in whom no metastases could be demonstrated at the time of testing [36]. Thus, although the predictive value of a negative result is high, i.e. a normal level of the high-molecular-mass fraction is unlikely to be found if metastases are present, the high proportion of apparent false positives in this series implied that the overall diagnostic efficiency of the test is not greater than that of measurement of total or liver-specific serum alkaline phosphatase. However, as in all such comparisons of diagnostic efficiencies, the possibility exists that the test being evaluated is more sensitive than the reference procedure, and that the apparently falsely elevated high-molecular-mass alkaline phosphatase levels in fact reflect the existence of undetected metastases.

Although high-molecular-mass alkaline phosphatase in serum is undoubtedly heterogeneous, it is almost always measured as a single entity. Further insights into the ways in which alkaline phosphatase gains access to the circulation, and perhaps increased diagnostic efficiency, can be gained from a consideration of the nature of the high-molecular-mass fraction.

High-molecular-mass alkaline phosphatase present in serum is frequently found to be associated with lipoprotein-X [38]. Since the latter is formed in plasma by combination of regurgitated lipid constituents of bile, mainly phospholipids and non-esterified cholesterol, with albumin and apolipoprotein-C, it can be assumed that alkaline phosphatase, solubilized or released from hepatocytes complete with its hydrophobic, membrane-binding domain, is able to participate in such complexes. It is probable that association of alkaline phosphatase with lipids takes place mainly in the bile, since biliary alkaline phosphatase occurs in high-molecular-mass complexes which can be dissociated by lipid-solvents, and that it is these regurgitated complexes that combine with lipoprotein-X in plasma (Fig. 5). Alkaline phosphatase released directly into plasma may also enter into complexes with lipids although this seems less likely: not only is hepatocyte alkaline phosphatase found in plasma in its lower-molecular-mass form in cholestasis (i.e., not complexed), but also the alkaline phosphatase deriving from osteoblasts in bone disease which, like hepatocyte alkaline phosphatase retains its membrane-binding domain on release into plasma [27], is present in a lower-molecular-mass form. The difference in complex-forming ability between plasma and biliary alkaline phosphatases may result from the retention of more of the CPI tail in the latter case, including perhaps the lipophilic acylglycerol moiety.

High molecular-mass forms of alkaline phosphatase that do not include lipoprotein-X also occur in serum (Fig. 5). Electron microscopy shows that these have a particulate structure and exhibit the bilayer structure of cell membranes [41]. As would be expected, these particles (also called "koinozymes") incorporate the activities of other enzymes also, such as γ-glutamyl transferase and 5′-nucleotidase, and in the series of patients with colorectal cancer already referred to, a strong positive correlation ($r = 0.89$) was found between high-molecular-mass forms of alkaline phosphatase and γ-glutamyl transferase [36]. Solubilization of alkaline phosphatase from the high molecular-mass fraction from serum with phospholipase C releases an enzyme with the electrophoretic properties of hepatocyte alkaline phosphatase, presumably by disrupting alkaline phosphatase-phosphatidylinositol interactions in hepatocyte membrane fragments [42]. As already noted, treatment of high-molecular-mass alkaline phosphatase from serum with bromelain releases alkaline phosphatase of the same molecular size as the enzyme solubilized from liver tissue by digestion with bromelain [27] (Fig. 6).

Increased levels of membrane fragments can be expected in destructive hepatocellular lesions, and are the main factor in determining the value of measurements of high-molecular-mass alkaline phosphatase in the detection of hepatic metastases.

5 Intestinal Alkaline Phosphatases and Hepatobiliary Disease

5.1 Adult-intestinal Alkaline Phosphatase

Adult-intestinal alkaline phosphatase is present in low activities (up to about 15% of total activity) in sera from about one third of all healthy individuals, more frequently in those who are of blood-groups B or O who are secretor-positive, and especially after a fatty meal [43]. The intestinal isoenzyme is readily detectable by its lower anodal mobility on serum zymograms (Fig. 2) than bone or liver phosphatases which is not altered by digestion with neuraminidase (due to its lack of terminal sialic acid residues). It also shares with placental alkaline phosphatase a sensitivity to uncompetitive inhibition by L-phenylalanine, but it differs from placental phosphatase in being reversibly adsorbed by immobilized Reactive Yellow 13 and it can be isolated from serum by affinity chromatography on this isoenzyme-specific ligand [44],

The presence of adult-intestinal alkaline phosphatase in serum is not markedly correlated with intestinal disease: e.g., in a series of 87 patients with various gastro-intestinal-tract diseases on whose sera alkaline phosphatase isoenzyme analysis was performed, only 22 displayed an intestinal phosphatase zone: an incidence no greater than in normal individuals. Raised levels in disease are more often observed in chronic liver disease than in intestinal disease, but the incidence of an increased serum intestinal alkaline phosphatase in disease is also subject to modification by the genetic trait that operates in healthy subjects.

Adult-intestinal alkaline phosphatase prepared from post-mortem tissue lacks the putative membrane-binding fragment that can be removed by treating other tissue alkaline phosphatases with bromelain in vitro [27]. However, the native isoenzyme

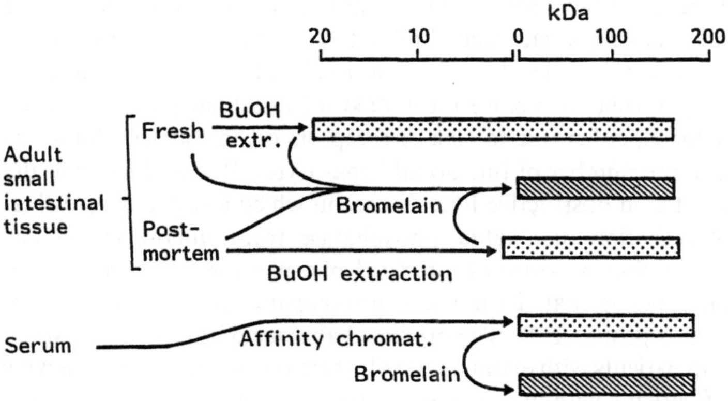

Fig. 7. Comparative molecular sizes determined by gradient-pore electrophoresis of various forms of human adult-intestinal alkaline phosphatase showing the effect of digestion with bromelain. Bromelain removes a fragment of approximately 21 kDa from the enzyme from fresh tissue, which is presumed to correspond to a hydrophobic, membrane-binding domain. This cannot be demonstrated in phosphatase extracted from post-mortem tissue, in which autolysis can be expected to have already taken place. Adult-intestinal alkaline phosphatase recovered from serum by specific affinity-chromatography on immobilized Reactive Yellow 13 similarly lacks the bromelain-cleavable, membrane-binding domain. (Note difference in scale to left of bromelain-cleavage point, 0)

does contain such a membrane-insertion domain as can be demonstrated with intestinal phosphatase extracted within 2 hours from fresh tissue obtained at operation, bromelain-treatment of which reduces its molecular weight by about 20 kDa (Fig. 7). The absence of the fragment removable by bromelain from post-mortem samples of intestinal alkaline phosphatase can be attributed to its removal by autolysis. Intestinal alkaline phosphatase recovered from patients' sera appears to be of similar size to the bromelain-treated isoenzyme from fresh tissue or the enzyme from post-mortem tissue, and it is not sensitive to digestion with bromelain (Fig. 7). Thus, it seems to lack the membrane-binding domain, unlike serum alkaline phosphatases of liver origin.

Adult-intestinal alkaline phosphatase gains entry to the circulation via the thoracic duct lymph, in which it can be demonstrated in large amounts after feeding [45]. Its molecular weight in serum suggests that release from the brush-borders of the enterocytes is accompanied by partial proteolysis. Since it is an asialoglycoprotein, adult-intestinal phosphatase is rapidly cleared from the plasma by hepatic glactosyl-glycoprotein receptors: thus it is this rapid clearance which normally maintains its level in plasma within narrow, genetically-influenced limits. In hepatobiliary diseases, as already discussed, increased production of alkaline phosphatase as a result of enzyme induction in the hepatocytes is followed by an increased rate of passage of the enzyme from these cells directly or indirectly into the plasma. The carbohydrate side-chains of hepatic alkaline phosphatase terminate in sialic acid residues which protect the enzyme from clearance by galactosyl receptors, and its half-life in plasma is of the order of 1–2 days, in contrast to less than 1 hour for the intestinal isoenzyme. Thus, the dominating influence on the level of liver phosphatase in plasma is its rate of entry into the circulation, in contrast to the dominating effect of rate of clearance in the case of adult-intestinal phosphatase.

This description of the routes of entry to and exit from the plasma of adult-intestinal alkaline phosphatase provide a basis for the interpretation of levels of this isoenzyme in disease. Its normally very rapid clearance limits its potential use as a marker of intestinal disease, so that raised levels are most likely to be due to impaired clearance. In severe hepatocellular disease the uptake of intestinal phosphatase by hepatocytes will be reduced, either due to lack of receptors or, in such conditions as cirrhosis, a reduction in the number of functional hepatocytes. Reduced entry to the circulation can be expected in obstructive liver disease in which absence of intestinal bile salts reduces solubilization of alkaline phosphatase from enterocytes. It has been suggested, therefore, that presence of intestinal phosphatase in the circulation of a jaundiced patient suggests that the lesion is intrahepatic, and its absence that extrahepatic obstruction is probable [46]. The predicted absence of intestinal phosphatase was confirmed in 15 patients with extrahepatic obstruction, whereas the isoenzyme was present in 45% of patients with intrahepatic lesions. A later study confirmed the presence of intestinal phosphatase in 42% of patients with intrahepatic jaundice, but in spite of a high specificity, the low sensitivity of the test resulted in an overall efficiency for the diagnosis of intrahepatic jaundice of only 57% [47].

5.2 *Foetal-intestinal Alkaline Phosphatase and the Kasahara Isoenzyme*

An alkaline phosphatase with the ligand-binding characteristics of the adult-intestinal isoenzyme is the predominant form in foetal intestine up to about 30 weeks gestation, declining in activity thereafter so that only the adult form is present at term [48]. Foetal-intestinal alkaline phosphatase is the major phosphatase of amniotic fluid at 18–20 weeks gestation [49], and is occasionally detectable in the serum of premature neonates [50]. Foetal intestinal alkaline phosphatase in serum is most easily distinguished from the adult isoenzyme by its sialic acid content, which gives it a rapid anodal electrophoretic mobility.

In spite of the many similarities between the alkaline phosphatases of foetal and adult small intestines, the foetal enzyme does not appear to be merely sialylated adult-intestinal phosphatase. The two forms are not identical in electrophoretic mobility after removal of sialic acid from the foetal enzyme, and certain monoclonal antibodies differentiate to some extent between the two forms [51]. The molecular weights of the two enzymes also differ, and they give rise to somewhat different peptide maps [52,53]. However, this latter evidence is perhaps inconclusive: alkaline phosphatase from post-mortem adult tissue was used in deriving the maps and, as has already been mentioned, enzyme from this source is already partially proteolysed whereas the foetal enzyme is not [27], presumably because of the immaturity of protease development in the foetus.

Although the structural evidence that foetal-intestinal alkaline phosphatase is a true, genetically-determined isoenzyme is therefore not yet conclusive, the existence of a separate structural gene encoding this enzyme seems probable. Selective expression of a specific gene could account for the limited developmental expression of foetal-intestinal phosphatase, while its ectopic re-expression could be the origin of the cancer-associated Kasahara isoenzyme. The Kasahara isoenzyme is a variant alkaline phosphatase occasionally detectable in the sera of cancer patients, especially but not exclusively those with hepatoma [54]. It has the inhibition characteristics of adult and foetal intestinal alkaline phosphatases and, like them, is bound to immobilized Reactive Yellow 13 [50]. Like foetal-intestinal alkaline phosphatase it is sialylated, but the molecular weight of one example was similar to that of the bromelain-treated, rather than the native, foetal enzyme [55]. Thus, the most similar normally-expressed analogue of the Kasahara isoenzyme is foetal-intestinal alkaline phosphatase, although the properties of the two are not always identical.

The Kasahara isoenzyme is one of several cancer-associated alkaline phosphatases of which the Regan isoenzyme has been studied most extensively. The appearance of the Regan isoenzyme in cancer is attributed to re-expression of the gene which encodes the term-placental alkaline phosphatase isoenzyme, with which is is virtually identical: by the same reasoning, therefore, the appearance of the Kasahara isoenzyme can be accounted for by the re-expression of a specific foetal-intestinal phosphatase gene. Differences between the normal foetal isoenzyme and the tumour-derived product could result from post-translational modifications of the latter, such as partial proteolysis or differences in glycosylation.

The Kasahara isoenzyme has been reported to be present in the plasma of 18%

of patients with hepatocellular cancer in Britain [56]. However, it is not specific for this condition, and its low incidence severely limits its diagnostic value. Like other tumour-derived alkaline phosphatases, its main value is in following the progress of patients in whom it is expressed and its identification helps in assigning a raised serum alkaline phosphatase to various tissue sources.

6 Concluding Remarks

The study of alkaline phosphatase in hepatobiliary disease provides an outstanding example of the way in which characterization of the multiple molecular forms of an enzyme can lead to insights into normal and pathological processes at the cellular and molecular levels which, in turn, can form the basis of new clinical interpretations. After more than 50 years of observation, experiment and debate, it can now be seen that, in their simple forms, neither the hepatogenic theory nor the retention theory is adequate fully to explain the changes in alkaline phosphatase activities and characteristics seen in hepatobiliary disease. Although increased synthesis of hepatic alkaline phosphatase — as required by the hepatogenic theory — is the primary response in biliary obstruction, elements of the retention theory are needed to explain the appearance of certain high-molecular-mass complexes of alkaline phosphatase ("biliary" alkaline phosphatase) in the circulation, by means of the regurgitation of bile. The latter probably occurs through increasingly permeable tight junctions between parenchymal cells, although whether this increasing permeability is such as to admit the passage of macromolecules remains somewhat controversial. Transport through cells by endocytosis at canalicular surfaces followed by exocytosis at sinusoidal surfaces may provide a concurrent, or alternative, pathway for alkaline phosphatase from bile to blood.

The liver does not appear to play the role in the removal from plasma of alkaline phosphatases from extra-hepatic sources, such as bone, envisaged in the retention theory, with the notable exception of alkaline phosphatase from the adult small intestine. The majority of intestinal alkaline phosphatase taken up by parenchymal cells via galactosyl-glycoprotein receptors appears to be catabolized within the cells. However, a small amount of intestinal alkaline phosphatase has been detected in hepatic bile, suggesting that at least part of the isoenzyme taken up by the hepatocytes is re-excreted [57]. The liver itself probably plays little part in clearing from plasma alkaline phosphatases from non-intestinal tissues such as bone, or released from parenchymal cells themselves. If these sialoproteins become de-sialylated, with the exposure of terminal galactosyl residues, they could then be taken up by parenchymal cells as is intestinal phosphatase. However, there is at present no evidence for or against this, and it seems more probable that sialylated alkaline phosphatases are removed from the circulation by tissue macrophages, amongst which the Kupffer cells, located in the liver, may be presumed to participate.

7 References

1. Roberts WM (1930) Br J Exp Path 11: 90
2. Polin SG, Spellberg MA, Teitelman L, Okumura M (1962) Gastroenterology 42: 431
3. Kaplan MM, Righetti A (1970) J Clin Invest 49: 508
4. Kryszewski AJ, Neale G, Whitfield JB, Moss DW (1973) Clin Chim Acta 47: 175
5. Sebesta DG, Bradshaw FJ, Prockop DJ (1964) Gastroenterology 47: 166
6. Hodson AW, Latner AL, Raine L (1962) Clin Chim Acta 7: 255
7. Moss DW, King EJ (1962) Biochem J 84: 192
8. Wootton AM, Neale G, Moss DW (1977) Clin Sci Mol Med 52: 585
9. Seetharam S, Sussman NL, Komoda T, Alpers DH (1986) Hepatology 6: 374
10. Hatoff DE, Hardison WGM (1982) Gastroenterology 77: 1062
11. Jersky J (1974) Gut 15: 636
12. Melani F, Farnararo M, Chiarugi VP (1971) Biochem J 121: 33
13. Riley PA, Spearman RIC (1968) Science 160: 1006
14. Haussler MR, Nagode LA, Rasmussen H (1970) Nature 228: 1199
15. Moog F, Etzler ME, Grey RD (1969) Ann NY Acad Sci 166: 447
16. Griffin MJ, Cox RP (1966) Proc Natl Acad Sci USA 56: 946
17. Muller E, Pearse AGE, Moss DW (1971) Biochem J 123: 895
18. Koyama I, Komoda T, Sakagishi Y, Kurata M (1983) Biochim Biophys Acta 760: 169
19. Schlaeger R, Haux P, Katterman R (1982) Enzyme 28: 3
20. Selden C, Wootton AM, Moss DW, Peters TJ (1978) Clin Sci Mol Med 55: 423
21. Blitzer BL, Boyer JL (1978) J Clin Invest 62: 1104
22. Hatoff DE, Toyota H, Wang C, Muller AL, Takeya M, Miyai K (1985) Dig Dis Sci 30: 564
23. Aronsen KF, Hagerstrand I, Norden JG (1972) Acta Path Microbiol Scand Sect A 80: 501
24. Moss DW, Whitaker KB (1985) Enzyme 34: 212
25. Haije WG, De Jong M (1963) Clin Chim Acta 8: 620
26. Moss DW, Campbell DM, Anagnostou-Kakaras E, King EJ (1961) Biochem J 81: 441
27. Kalimanovska V, Whitaker KB, Moss DW (1987) Clin Chim Acta 170: 219
28. Moss DW, Edwards RK (1984) Clin Chim Acta 143: 177
29. Rosalki SB, Foo AY (1984) Clin Chem 30: 1182
30. Peacock AC, Reed AC, Highsmith EM (1963) Clin Chim Acta 8: 914
31. Moss DW, Shakespeare MJ, Thomas DM (1972) Clin Chim Acta 40: 35
32. Whitby LG, Moss DW (1975) Clin Chim Acta 59: 361
33. Moss DW, Whitby, LG (1975) Clin Chim Acta 61: 63
34. Whitaker KB, Whitby LG, Moss DW (1977) Clin Chim Acta 80: 209
35. Crofton PM, Elton RA, Smith AF (1979) Clin Chim Acta 98: 263
36. Traynor OJ, Wood CB, Echetebu ZO, Whitaker KB, Moss DW (1986) Br J Cancer 53: 483
37. Crofton PM, Smith AF (1981) Clin Chem 27: 860
38. Brocklehurst D, Wilde CE, Doar JWH (1978) Clin Chim Acta 88: 509
39. Siede WH, Seiffert UB (1983) Clin Chem 29: 698
40. Viot M, Joulin C, Cambon P, Krebs BP, Schneider M, Lalanne CM (1979) Biomedicine 31: 74
41. De Broe ME, Borgers M, Wieme RJ (1975) Clin Chim Acta 59: 369
42. Sykes E, Kiechle FL, Epstein E (1986) Clin Chem 32: 1503
43. Langman MJS, Leuthold E, Robson EB, Harris J, Luffman JE, Harris H (1966) Nature 212: 41
44. Williams DG, Byfield PGH, Moss DW (1982) Enzyme 28: 28
45. Keiding NR (1964) Clin Sci 26: 291
46. Warnes TW, Hine P, Kay G, Smith A (1977) Gut 18: 274
47. Rosalki SB (1986) Adv Clin Enzymol 3: 12
48. Mulivor RA, Hannig VL, Harris H (1978) Proc Natl Acad Sci USA 75: 3909
49. Mulivor RA, Mennuti MT, Harris H (1979) Am J Obstet Gynecol 135: 77
50. Moss DW, Parmar CR, Whitaker KB (1986) Clin Chim Acta 158: 165

51. Vockley J, Meyer LJ, Harris H (1984) Am J Hum Genet 36: 987
52. Vockley J, D'Souza MP, Foster CJ, Harris H (1984) Proc Natl Acad Sci USA 81: 6120
53. Mueller HD, Leung H, Stinson RA (1985) Biochem Biophys Res Comm 126: 427
54. Higashino K, Otani R, Kudo S, Yamamura Y (1977) Clin Chem 23: 1615
55. Moss DW, Whitaker KB (1987) Clin Biochem 20: 9
56. Kay P, Johnson PJ, Warnes TW, Williams R, Longson M, Laing I, Smith A, Lee NW (1982) Gut 23: A 907
57. Warnes TW, Hine P, Kay G, Smith A (1981) Gut 22: 493
58. Ferguson MA, Williams AF (1988) Ann Rev Biochem 57: 285

Enzymatic Profiles of Hepatic Disease

Investigated by Alkaline Phosphatase Isoenzymes and Isoforms

John Griffiths

Medical University of South Carolina, Department of Pathology/Laboratory Medicine Charleston, SC 29425, USA

Current zonal electrophoretic techniques for the separation of alkaline phosphatase isoenzymes are unsatisfactory for "in depth" clinical use. A recently described isoelectric focusing technique using agarose gel containing both carrier ampholytes and separators over a pI range of 3.01 to 4.86, suggests at least twelve bands of ALP activity. Four of the most acidic bands appear in serum, with activity above reference range, in the presence of clinically confirmed hepatic disease. It is thought that each band represents ALP secreted by a distinct hepatic cellular architecture. Various hepatic diseases are illustrated by patient profiles, from which a pattern recognition developed. It is emphasized that the patterns are reflections of serum ALP activity, without pre-treatment.

The literature is examined for previous experience with ALP isoenzymes in the confirmation of a differing concept for hepatic disease — the biochemical biopsy.

Progress in Clinical Biochemistry
and Medicine, Vol. 8

1 Introduction

The group of membrane located enzymes collectively termed the alkaline phosphatases (E.C. 3.2.1.1., ALP) are said to originate in 3 genetically distinct sites. Intestine, placenta, present in the pregnant female and a somewhat ambiguous tissue non-specific [1]. The latter, by definition may infer several differing points of cellular origin. This suggestion, if correct would indicate three bands of activity in serum when traditional methods of zonal electrophoresis are used. In practice, the serum of the ostensibly healthy and the pregnant patient, shows 2 to 3 bands, bone — liver, placental and occasionally intestinal. These are presently designated ALP isoenzymes. Despite interesting variations to enhance zonal electrophoresis such as heat inhibition [2], treatment with neuraminidase [3] and lectin binding [4], no laboratorian can be satisfied with contemporary techniques.

2 Method

Recently, an isoelectric focusing system (ISEF) utilizing agarose gel, containing carrier ampholytes and separators to achieve a pI range of 3.01 to 4.86 has revealed at least twelve bands of varying ALP activity in normal plasma (Table 1) [5]. The nine additional bands may be considered isoforms of the original isoenzymes — that is post translational modifications, formed in individual groups of specialized cells or in plasma. The widespread histo-pathological distribution of ALP in tissues infers that some isoforms may be individually derived in the cells of those organs.

3 Main Clinical Issue

The major interest of the clinician when confronted with elevated total activity of ALP is to identify the tissue source. Practically, origination from bone or liver is the most frequent question. ISEF is able to make a clear distinction with the separation of Band 5 at a pI of 3.65, (of bone origin) from Bands 1 to 4 which extend over a more acidic range. Experience has now indicated that these ALP bands vary greatly in concentration with a plethora of liver diseases and their distinctive appearance indicates a specific profile or pattern.

4 Approach

It would be naive to consider the liver as a single group of cells. The 2000 gram total mass of liver contains at least 250 grams of reticulo endothelial tissue, 250 grams of vascular bed (sinusoid) and at least 350 grams of biliary tract [6]. Such cell concentrations are surely capable of constructing their own cellular messages such as ALP.

Table 1. Alkaline phosphatase isoenzymes/isoforms in healthy subjects according to age

ALP band no.	Isoelectric point, pI	Tissue source	% of total ALP, range				
			<1 yr[a] (n = 247)	1–15 yr (n = 291)	Adults[b] (n = 1426)	Post-menopausal women (n = 309)	Pregnant women[c] (n = 260)
1	3.01	Biliary canaliculus	3–6	2–5	1–3	0–12	1–3
2	3.65	Hepatocyte	13–20	14–20	10–20	10–34	2–10
3	3.88	Hepatic reticuloendothelial cells	7–14	8–14	7–15	7–14	3–7
4	4.10	Vascular bed, <25 μm	10–18	9–15	8–12	6–9	3–8
5	4.21	Bone	20–30	21–30	13–19	8–21	8–14
6	4.35	Pulmonary tissue	3–8	4–9	3–10	4–8	6–9
7	4.45	Neutrophil	6–12	6–14	4–8	2–8	2–5
8	4.55	Fetal-tissue-derived	4–8	4–9	4–13	5–9	5–17
9	4.65	Placental (late)	4–9	4–9	5–11	3–5	18–25
10	4.73	Placental (early) or activated T lymphocyte	4–10	1–8	8–10	4–10	35–44
11	4.80	Renal	1–3	0–1	0–2	0–2	3–6
12	4.86	Intestinal	0–2	0–1	0–1	0–1	0–1
Range of total ALP acty, U/L			190–310	160–240	40–110	30–130	170–230

[a] Age range: 1 day to 1 year postpartum. [b] Age range: 16–50 years. [c] Third trimester.

Other structures like lymphatics, Ito cells and fat cells may conversely not have sufficient generative capacity to reach adequate concentrations in a 5 liter blood pool.

In observing varying patterns of activity in Bands 1 through 4, and applying this recognition to disease, a classification of hepatic disease is helpful. In general, a simplistic one such as outlined in Table 2, is sufficient for everyday use. It is, however, stressed that no cellular group exists independently in the complex mechanism of hepatic function and that cellular turmoil in one group may inevitably be reflected in other groups, albeit at varying temporal intervals.

Table 2. Simplistic classification of hepatic disease

Architecture	Disease — example
1. Canalicular	Primary biliary cirrhosis
2. Hepatocyte	Acute, sub acute and chronic hepatitis
3. Vasculature (small vessels) <25 micron	Multiple cyst disease
4. Mixed	Auto immune discordant ALP syndrome

Table 3. Laboratory upper reference ranges for relevant tests

Total protein	7 G/dL
Albumin	3–4 G/dL
Prothrombin time	over 12 secs.
Aspartate amino transferase (AST)	35 U/L
Alanine amino transferase (ALT)	30 U/L
Alkaline phosphatase (ALP)	110 U/L

5 Patient Profiles

This paper is illustrated by a series of eight case reports selected to encompass our simple classification of liver diseases.

5.1 Canalicular Origin

A 47 year female, asymptomatic, was found on routine biochemical investigation to show a total ALP activity of 800 U/L. ISEF fractionation revealed a marked increase of Band 1 at 329 U/L, and of Band 2 at 276 U/L. The remaining bands, both 3 and 4, from the liver and from described other sources are within reference. The profile suggests the major band of activity to be biliary in origin. Percutaneous liver biopsy confirmed primary biliary cirrhosis as the diagnosis (Fig. 1).

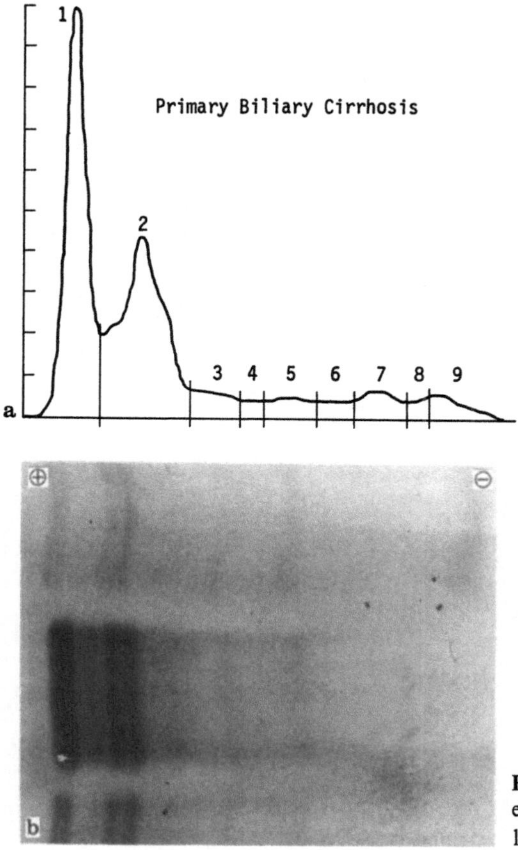

FRACTION	%	U/L
1	41.19	329.53
2	34.57	276.55
3	4.98	39.87
4	1.64	13.11
5	4.00	32.03
6	2.77	22.19
7	4.79	38.36
8	1.59	12.69
9	4.46	35.67
TOTAL		800.00

Fig. 1. Primary biliary cirrhosis: Indicates elevation of Band 1 activity (Canaliculus) and lesser elevation of Band 2 (Hepatocyte)

5.2 Mixed Cellular Architecture

A 24 year female in the 37th week of pregnancy was found to have a mild elevation of total bilirubin at 1.9 mg/dL — of which 1.2 mg/dL was albumin bound and 0.7 mg/dL, glucuronide bound. Total ALP was 190 U/L with 230 U/L accepted as upper limit of reference for this stage of pregnancy. ISEF indicated elevation of Bands 1 and 2 indicating mild cholestatic hepatocyte involvement as well as elevation of Bands 8, 9 and 10 associated with pregnancy. Other laboratory parameters of hepatic function were within reference. The pregnancy proceeded to a normal conclusion and within 5 days, all bands were within reference limits — diagnosis mild cholestatic jaundice of pregnancy (Fig. 2).

5.3 Hepatocyte Dominant

23 year male, known to be an I.V. drug addict presents with an acute febrile illness of 3 days duration. Laboratory data showed mild increase in total bilirubin 1.4 mg/dL, ALT — 45 U/L (reference to 30 U/dL) and extension of prothrombin time. Other

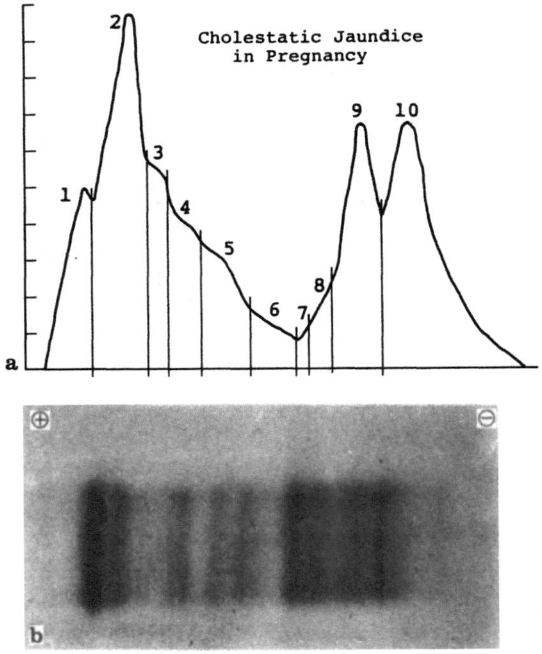

FRACTION	%	U/L
1	9.26	17.59
2	24.24	45.05
3	5.95	11.30
4	7.90	15.01
5	8.36	15.88
6	3.23	6.13
7	0.70	1.34
8	2.81	5.34
9	14.99	28.49
10	22.57	42.88
TOTAL		190.00

Fig. 2. Cholestatic jaundice in pregnancy: Elevation in ALP activity of Band 1 (Canalicular) and Band 2 (Hepatocyte). There is also elevation of Bands 8, 9 and 10 associated with pregnancy

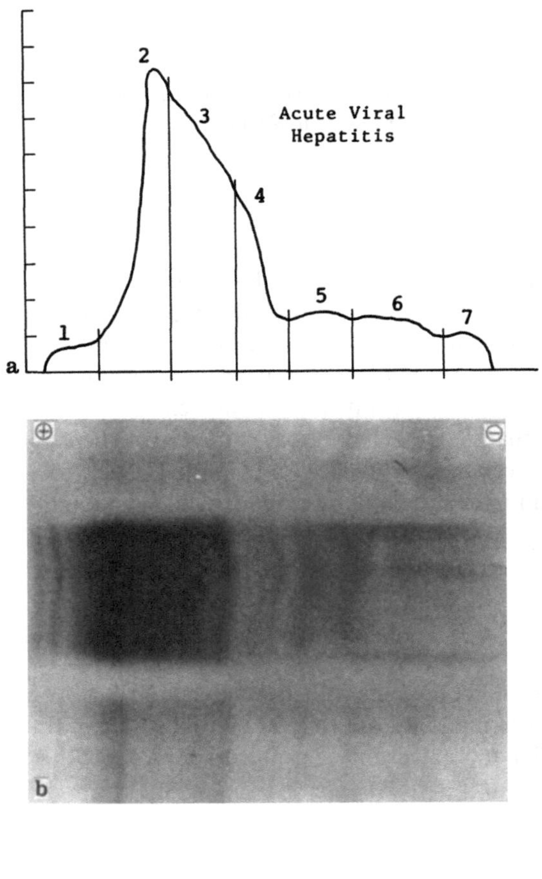

FRACTION	%	U/L
1	3.42	3.76
2	24.75	27.22
3	37.29	41.02
4	11.68	12.85
5	8.56	9.41
6	11.10	12.21
7	3.20	3.52
TOTAL		110.00

Fig. 3. Acute viral hepatitis: Major ALP activity is noted in Band 2 (Hepatocyte) and Band 3 (Mono-histiocyte liver)

parameters such as total protein, albumin and AST within reference, while ALP was 110 U/L. The fractionation showed elevation in Bands 2 and 3 with Bands 1 and 4 within references. A liver biopsy at this stage suggested acute viral hepatitis confirmed by subsequent immunological studies for Hepatitis A virus (Fig. 3).

5.4 Hepatocyte Dominant

37 year female known to have viral hepatitis (Type A) 3 years previously, whose continued follow-up had indicated variable abnormal laboratory data. At the time of investigation, she showed hypoalbuminemia (3.0 G/dL) moderate elevation of AST (39 U/L) and ALT (72 U/L) and a total bilirubin of 1.6 mg/dL. Total ALP was 160 U/L. ISEF showed elevations of all hepatic ALP components but Band 2 was dominant. Biopsy indicated continuation of chronic active hepatitis (Fig. 4).

5.5 Hepatic Mixed

Male of 57 years with a long history of excessive alcohol intake. His presentation with bleeding esophageal varices and the subsequent physical examination strongly suggested hepatic cirrhosis. Total protein 6.2 G/dL, albumin 1.2 G/dL. Total bilirubin

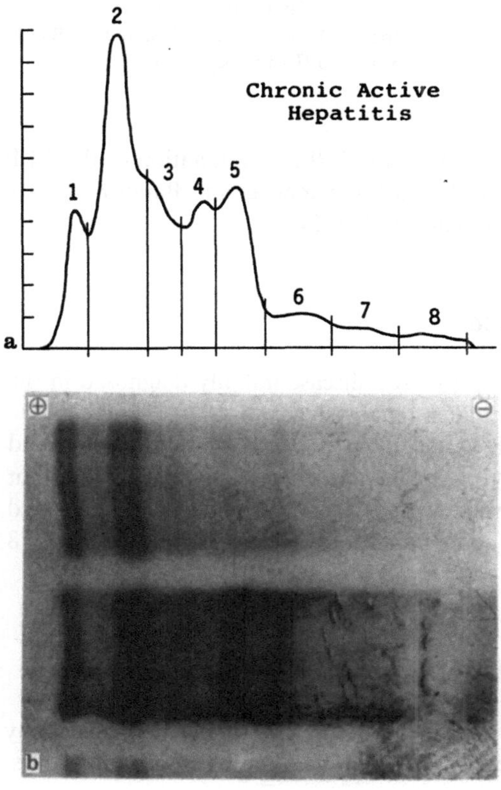

FRACTION	%	U/L
1	8.80	14.08
2	35.59	56.95
3	15.02	24.03
4	12.85	20.56
5	15.73	25.17
6	6.15	9.84
7	3.68	5.88
8	2.18	3.48
TOTAL		160.00

Fig. 4. Chronic active hepatitis: Slight elevation of activity in Band 1 (Canaliculus) and major elevation of Band 2 (Hepatocyte). Slight elevation of Band 3 (Mono-histocyte liver) and Band 4 (Small Vessel — sinusoid liver)

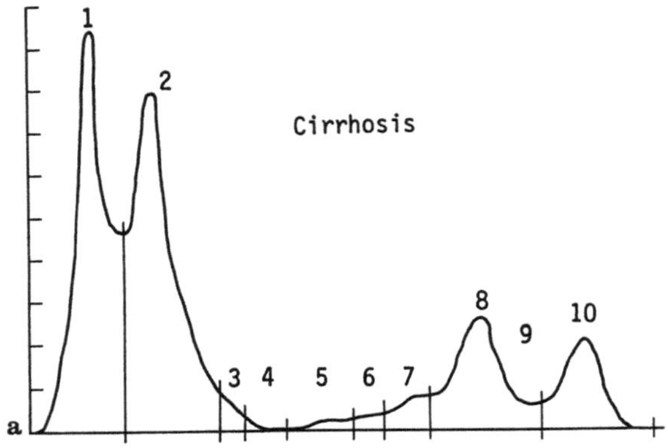

FRACTION	%	U/L
1	34.80	55.69
2	34.99	55.99
3	1.23	1.96
4	0.48	0.77
5	0.92	1.48
6	0.99	1.59
7	2.80	4.49
8	14.46	23.14
9	9.31	14.90
TOTAL		160.00

Fig. 5. Hepatic cirrhosis: Elevation of Band 1 (Canaliculus) and Band 2 (Hepatocyte). Elevation of ALP activity in Bands 8 and 10 (Lymphocyte derived)

2.4 mg/dL ALT 24 U/L, AST 32 U/L and ALP 160 U/L. Fractionation of the ALP showed marked increases in Bands 1 and 2, and lesser increases in Bands 8 and 10. Hepatic biopsy confirmed the cirrhotic diagnosis (Fig. 5).

5.6 Possible Vascular Dominance

53 year female known to have multiple polycystic disease initially diagnosed in the kidney, with onset of renal failure and later renal osteodystrophy. Her creatinine clearance was 21 mL/minute and ALP varied from 372 U/L to 120 U/L pre and post dialysis. Multiple angiomatous cystic areas were noted in liver and lung but other parameters of hepatic function were within reference limits. ISEF indicated increases in Bands 2, 3 and 4. The bone Band 5 was elevated as was Band 8 (Fig. 6).

5.7 Possible Hepatocyte Dominance

Level ALP discordant to level total bilirubin. 74 year female with post hemi-colectomy for carcinoma of colon. Post operative difficulties in maintaining nutritional status,

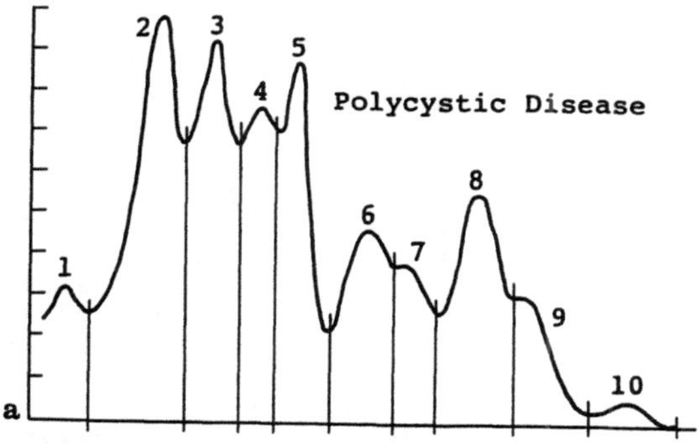

FRACTION	%	U/L
1	5.38	20.01
2	22.31	83.01
3	16.95	63.05
4	9.25	34.41
5	12.87	47.88
6	9.23	34.34
7	5.65	21.01
8	12.04	44.77
9	5.10	18.97
10	1.22	4.55
TOTAL		372.00

Fig. 6. Multiple polycystic disease: Elevation of ALP activity in Band 2 (Hepatocyte) Band 3 (Monohistiocyte liver) Band 4 (Small vessel) Band 5 (Bone) Band 6 (Pulmonary) and Band 8 (Lymphocyte derived)

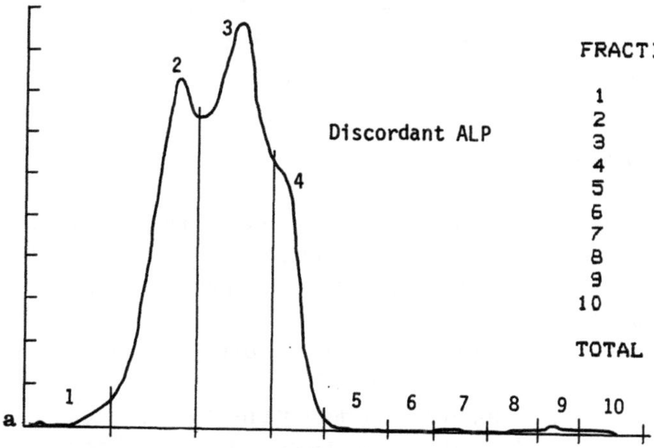

FRACTION	%	U/L
1	1.25	15.05
2	35.04	420.42
3	47.60	571.17
4	14.05	168.56
5	0.42	4.98
6	0.12	1.44
7	0.34	4.09
8	0.36	4.28
9	0.50	5.95
10	0.34	4.07
TOTAL		1200.00

Fig. 7. Discordant: Marked elevation of Bands 2, 3 and 4 suggesting enzyme induction in the liver

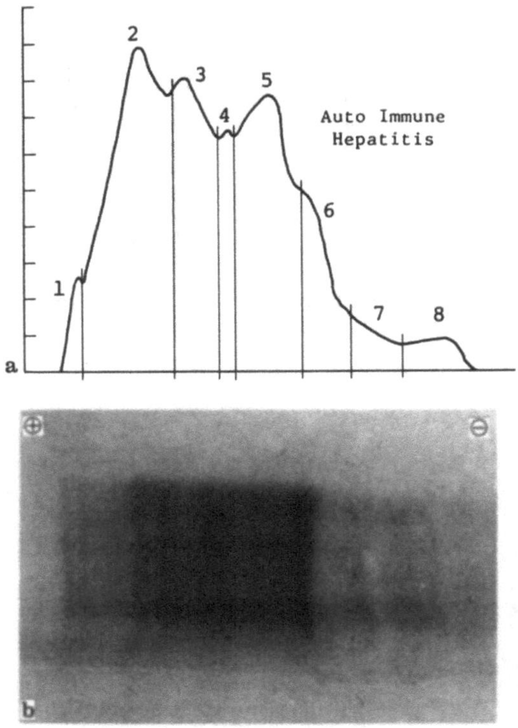

FRACTION	%	U/L
1	2.51	22.71
2	33.17	300.14
3	19.33	174.92
4	5.13	46.47
5	25.55	231.27
6	8.76	79.32
7	3.13	28.31
8	2.41	21.35
TOTAL		905.00

Fig. 8. Auto immune hepatitis: Elevation of Band 2 (Hepatocyte) Band 3 (Monohistiocyte) and of Band 5 (Bone) in a young child

necessitated hyperalimentation. There was hypoalbuminemia of 2.1 G/dL, total bilirubin was 1.2 mg/dL, ALT and AST, gamma GT, within reference. ALP varied between 1100 to 1200 U/L. ISEF showed elevation in Bands 2, 3, and 4 (Fig. 7).

5.8 Hepatocyte Dominance

Two year old female with two month history of vague abdominal pain, occasional loose bowel motion and a fleeting arthralgia. Physical examination not diagnostically suggestive. In laboratory investigations, the only abnormal findings were, antinuclear antibody, titer 1:256 (speckled pattern) and total ALP — 905 U/L. ISEF showed an increase in Bands 2 and 3, while Band 5, the bone component was elevated but compatible with the childs age. In liver biopsy, the basic H and E stains were not diagnostically helpful. Fluorescent antibody stains, however, indicated heavy deposits of IgG, IgA but not IgM in the sinusoidal cells, as well as fibrinogen and complement C3. The diagnosis of auto-immune hepatitis was formulated (Fig. 8).

6 Concluding Remarks

The concepts expressed in this chapter depend on the realization that the liver is composed of a heterogenous population of cells. No longer can we utilize the broad term 'liver' since each group of cells that compose that anatomical structure are capable of generating individual biochemical compounds (signals) relaying these to the vascular space.

If the signals assayed are enzymes, then present techniques such as isoelectric focusing allow the differentiation of such enzymatic compounds by their activity on appropriate substrate; after specific separation, by pI gradients.

Recent studies with creatine kinase 3 isoforms suggest each is different by a single amino acid (lysine) in a carboxypeptidase A removal from the parent isoenzyme [7]. This observation infers excellent sensitivity of ISEF, in distinguishing small molecular differences. This has not been possible in the hybridoma system.

The conclusions in this presentation are based on the cardinal observation that ALP changes are seen in the serum of patients with histologically documented, hence specific clinical hepatic disease. No treatment by heat, neuraminidase or other means has been introduced into the technique. It is noted that ALP tissue enzymology also studied by electrophoresis may not compare precisely with serum findings [8].

One of the four ALP bands of activity is probably the basic isoenzyme (tissue non specific). The remaining bands may be termed isoforms. However, it is considered that they are derived from different functional cells within the liver like hepatocyte and canaliculus, thus not post translational. The idea may yet be 'not proven', without knowledge of structural form. The comparison to other known isoforms derived from creatine kinase isoenzymes 3 and 2 may then be premature.

Sequentially examining each ALP band of activity, Band 1 at pI 3.01 appears to reflect clinical disturbance of canalicular function, observed in a number of disease entities where the biliary tract is either primarily or secondarily involved. Destruction with rearrangement, induction by bile acid pooling or by excretion of chemical catabolic products produce an unmistakable signal in the serum.

The magnitude of response varies according to pathology, the most intense, in primary biliary cirrhosis. The largest increases in ALP activity are seen in two situations, with obstructive canalicular disease, and the discrepant bilirubin/ALP syndrome. In the former, the ALP on the canalicular membrane may be solubilized by bile acids accumulating in the stasis, then enter the canalicular cells and further into the vascular spaces. In the latter, increased synthesis may be due to enhanced translation of mRNA. Band I of ALP activity may correspond to the 'fast pre-liver' zone described by Fritsche [9]. Where does the large MW liver cell membrane fragment conjugate noted by Epstein et al. [10] in cholestatic disease interrelate with Band 1? These authors summarize a wide range of experience also reflected by Crofton [11]. This zone of ALP activity presumably linked with large protein molecules is said to have little electrophoretic mobility, remaining at the point of application in zonal electrophoresis techniques. No such phenomenon has been noted in isoelectric focusing and the clinical significance of this observation remains unclear.

Band ALP 2 activity which focuses at pI 3.65 increases when the hepatocyte is affected by lethal destructive disease (viral or chemical in origin), by late cholestatic or obstructive disease — possibly similar to the mechanisms previously described, or by biochemical turmoil from non-lethal sources like drug induction. An example cyclosporin/steroid combination, precisely, hepatocyte located ALP has not been previously mentioned in hepatic diseases and correlated specifically with such diseases. The suggestion for such ALP location is derived from the enzyme mapping plan developed by Reith [12]. This indicated concentrations of ALP at the perivenous (hepato-vascular) and peri-canalicular (hepato-canalicular) apposition zones. Any pathological reason for cellular upheaval would be reflected in an easy egress of ALP from the hepatocyte into the vascular space. Histological evidence illustrates distinct hepatocyte changes in advanced biliary cirrhosis, acute and chronic active hepatitis and in active hepato destructive episodes of cirrhosis.

The band that locates at pI 3.88 and is considered to represent an ALP signal from the mono-histocyte bed (reticulo-endothelial or Kupffer Cell) is speculative at this point. No disease involving these cells singly is recognized, but certain observations point in that direction. In practice, this band was noted, solely increased in activity, was in cases of erythromycin idiosyncrasy where elevations of total ALP rose 5-7-fold in magnitude, regressing on drug withdrawal. The single reference support is made by Peticlerc [13] who considered the majority of circulating ALP to be derived from the general reticulo-endothelial system.

Increase in ALP activity relative to Band 4, pI 4.10 occurs in almost all hepatic diseases although the magnitude may vary. The suspicion of the ALP origin is the small vascular (sinusoid) bed — and may not solely be confined to liver vessels but to other small vascular areas. Corroborative histochemical evidence indicates high concentrations of ALP in the endothelial cells of terminal arterioles less than 25 micron in internal diameter and in capillary endothelium. There is no significant ALP in large blood vessels [14]. In clinical practice, pure hepatic vascular disease is rare, however the described case of multiple hepatic angiomatosis may add some credence to the suggestion. Information can only be accumulated by continued case studies.

This preliminary report has postulated that there are at least four distinct ALP isoenzymes/isoforms located in specific hepatic cells. The many hepatic diseases may release varying magnitudes of such ALP signals. These will present in differing patterns since no hepatic disease is monocellular.

The art of using the technique of isoelectric focusing in the evaluation of hepatic derived ALP will revolve around a pattern recognition. This inexpensive, non-invasive technique is a useful adjunct as a biochemical biopsy to the more traditional needle technique.

7 References

1. Stigbrand T (1983) Human alkaline phosphatases 3. Alan R Liss, Inc, New York
2. Moss DW, King EJ (1962) Biochem J 84: 192
3. Schreiber WE, Sadro LC (1988) Amer J Clin Path 90: 181
4. Onica D, Sundblad L, Waldenlind L (1986) Clin Chim Acta 155: 285
5. Griffiths J, Black J (1987) Clin Chem 33: 2175
6. O'Brien MJ, Gottlieb LS (1979) Pathological basis of disease, WB Saunder, London
7. Perryman MB, Knell JD, Roberts R (1984) Clin Chem 30: 662
8. Teder M, Parik J (1985) Hum Genet 71: 215
9. Fritsche HA, Adams Park HR (1972) Clin Chem 18: 417
10. Epstein E, Kiechle FL, Zak B (1984) Annals Clin Lab Sci 14: 292
11. Crofton PM (1982) CRC Rev in Clin Lab Sci Mar 161
12. Reith A (1979) in: Schmidt E, Schmidt FW (eds) Adv Clin Enzym 53, Karger, Basel
13. Petitclerc C (1976) Clin Chem 22: 42
14. Romanul FCA, Banister RG (1962) J Cell Biol 15: 73

Reference Values and Drug Effects on Hepatic Enzymes

Y. Artur[1,2], G. Siest[1,2], G. J. Sanderink[1], M. Wellman[1,2], M. M. Galteau[1,2], F. Schiele[1]

[1] Laboratoire du Centre de Médecine Préventine, UA CNRS° 597, 2 avenue du Doyen J. Parisot, 54500 Vandoeuvre-les-Nancy, France
[2] Centre du Médicament, 30 rue Lionnois, 54000 Nancy, France

Aminotransferases, alkaline phosphatase, gamma-glutamyl transferase and, to a lesser degree, alanine aminopeptidase are the most currently used enzymes for diagnosis and monitoring of hepatobiliary diseases. For a correct interpretation of activity measurements, clinicians need to dispose of reference values. Their production requires a good knowledge of the biological sources of variations affecting the enzymes and of the effects of intake of alcohol, tobacco, or drugs. The influence of some environmental factors has also to be taken into account. After a review of the main biological factors of variation affecting the enzyme activities, such as age, sex, morphometry, hormonal status, etc., the text focuses on the biological effects of anticonvulsants, hypolipidemic agents, and oral contraceptives on the liver enzymes in serum, excluding the toxicological aspects. The effects of alcohol and tobacco are also summarized.

Progress in Clinical Biochemistry
and Medicine, Vol. 8

1 Hepatic Enzymes in Clinical Chemistry

Aspartate aminotransferase (AST, EC 2.6.1.1), alanine aminotransferase (ALT, EC 2.6.1.2), alkaline phosphatase (ALP, EC 3.1.3.1), gamma-glutamyl transferase (GGT, EC 2.3.2.2) and, to a lesser degree, alanine aminopeptidase (AAP, EC 3.4.11.2) are the most currently used enzymes for diagnosis and monitoring of hepatobiliary diseases.

ALT and AST are present in cytosol and mitochondria. Both these enzymes are markers of cytolysis, their serum activities being largely increased in cases of liver cell necrosis (hepatitis, etc.).

ALP, GGT and AAP are localized in liver cell plasma membranes [1]. In normal liver, ALP is situated at the sinusoidal surfaces of the centrolobular vein and portal tract area. GGT is present in the sinusoidal domain, and to a lesser extent in the cell membranes of large bile ducts. Conversely, AAP is mainly located in the canalicular membranes. Serum activities of these three enzymes are increased in cholestasis; in this case, their distribution in liver plasma membranes is modified. Besides its use as a marker of cholestasis, serum GGT is an indirect index of enzyme induction by xenobiotics, especially alcohol and drugs. This is particularly of interest in monitoring of detoxication of alcoholics.

Four of these five enzymes are known to exist in serum in multiple molecular forms. The presence of true isoenzymes encoded by distinct structural genes has been demonstrated for AST [2] and ALP [3]. Conversely, post-traductional modifications of a single gene product are responsible for the heterogeneity of GGT [4], AAP [5], and also ALP [6] besides the true isoenzymes.

Serum ALT, GGT and AAP are known to originate mainly from the liver, whereas AST and ALP activities in serum are due to the presence of isoforms originating from various organs: liver and heart, principally, for AST, liver, bone, intestine and placenta for ALP.

For a correct interpretation of activity measurements, clinicians need to dispose of reference values. Their production requires a good knowledge of the analytical and biological sources of variation affecting the enzymes and of the effects of intake of alcohol, tabacco or drugs. The influence of some environmental factors has also to be taken into account.

This review will dwell especially on the physiological factors of variation of the five serum enzymes enumerated above, and on the effects of drugs on their activities.

2 Reference Values

2.1 *Use of the Concept of Reference Values* [7]

Reference values are values measured in healthy individuals selected according to well-defined criteria, especially taking into account the potential factors of variation;

these values allow the interpretation of the test results as a function of the objectives. Reference values are most frequently presented as reference intervals. The general convention is that the reference interval contains a proportion equal to 95% of the population, excluding 2.5% from either side of the distribution.

Clinicians need to dispose of a series of reference distributions in order to fix decision limits appropriate for each particular situation, e.g., verification of state of health, alerting the patient to current risks, confirmation of a diagnosis, making a therapeutic decision, monitoring of a therapy, etc. A value measured in a given person must be interpreted in its proper physiological and clinical context. Reference values can be used from two independent but complementary approaches: comparison of reference values with each other (comparison of two populations) and comparison of one isolated value with the reference values.

2.1.1 Comparison of Reference Values with Each Other

This approach especially concerns the use of reference values in epidemiology and anthropology, and in the study of drug effects:

- Epidemiology, anthropology: Reference values allow to compare ethnic behaviours, nutritional and social-cultural habits, and genetic factors. Long-term changes in the state of health of a population can also be studied.
- Study of drug effects: the effects of drugs on laboratory tests may be determined by comparing the distributions obtained from subjects undergoing treatment and for controls.

2.1.2 Comparison of an Observed Value with Reference Values

This is the most common use of reference values. The three main examples of this use concern health maintenance, interpretation of laboratory test results, and monitoring of therapies:

- Health maintenance: Reference values are of interest for detection of diseases before clinical signs become apparent, for classification of subjects in "at risk" groups, for evaluation of the probability of appearance of a disease, for comparison of successive health screenings, etc.
- Interpretation of laboratory test results in diseased patients: it represents the traditional way of interpretation of laboratory results. The observed values were formerly compared to "normal values".
- Monitoring of therapies: this use concerns the evaluation of a therapeutic effect, the monitoring of a risk related to a treatment or of individual effects such as enzyme induction by drugs.

2.2 Production of Reference Values [7,8]

2.2.1 Strategy for Establishing Reference Values

The strategy for producing reference values is now well established. A well-defined protocol must be followed, requiring the cooperation of clinicians, clinical chemists, technicians, statisticians and epidemiologists. Briefly, it is necessary to know the sources of variation of the laboratory test, to decide on the exclusion factors for eliminating from the reference sample subjects who are sick or who show some bias relative to the reference population, and to choose stratification factors, or partition criteria, in order to obtain homogeneous subpopulations. After preparing the subjects for blood sampling, blood is collected and analyses are performed under well-defined conditions. The last step consists of a statistical treatment of data. Knowledge of the variation factors affecting laboratory tests is an important step for the production of reference values. These sources of variation are summarized in the next paragraph.

2.2.2 Sources of Variation

Figure 1 gives a classification of the sources of analytical and biological variation. The total observed variation is the sum of two components: analytical variation and biological variation. Analytical variation is largely dependent on the method of measurement, whereas biological variation depends on the subject being studied. Biological variation may be divided into pathological and physiological variation. However, only physiological variation is studied in the present review.

- Analytical variability (preinstrumental and instrumental variation).
  Preinstrumental sources of variation include those introduced during blood collection, e.g., the effect of posture and use of a tourniquet and all other factors up to the introduction of the sample into the analytical system. Instrumental variation is attributable to factors in the measurement process itself.
- Biological variability.
  Biological variation can be derived from a single subject (intra-individual variation) or from the variation between subjects (interindividual variation).
  Intraindividual variation comprises all the influences affecting the composition of body fluids. Such influences include biological rhythms, aging and nutritional status. It is difficult to isolate separate sources of variation and to study them independently, since they are all closely related.
  Interindividual variation is determined from studying population groups. It includes the influence of genetic differences as well as temporal changes due to environment and other exogenous and endogenous factors. Table 1 lists the main biological factors that influence the results of laboratory tests.

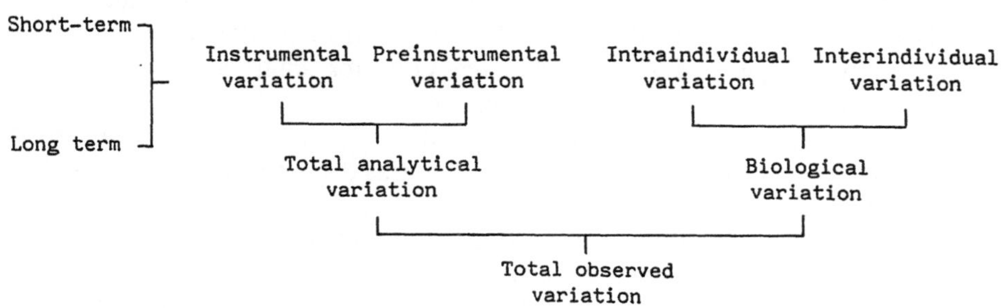

Fig. 1. Sources of variation of a laboratory test

Table 1. The main biological factors of variation

• *Age*	• Muscular exercise
• Altitude	• Nutrition
• Bed rest	• Pregnancy
• Blood pressure	• Puberty
• *Environmental factors* (including drugs, alcohol, tabacco)	• Rhythms
• Fasting	• *Sex*
• Genetic factors (race, blood group, . . .)	• Social-professional status
• *Menopause*	• Stress
• Menstrual cycle	• *Weight*

Analytical variations affecting the activities of serum AST, ALT, ALP and GGT have been reviewed elsewhere [9,10,11,12]. There are only a few data concerning analytical variation of serum AAP [13,14]. However, analytical variations of the enzymes are not developed here, and the next paragraph is devoted to the biological variations of the five enzymes.

2.3 Biological Variations of Hepatic Enzymes in Serum

Biological factors of variation of serum AST, ALT, ALP and GGT activities have been studied in detail [9,10,11,12]. The aim of this paragraph is to offer a synthetic overview of these factors, including recent data, especially in the case of AAP.

2.3.1 Age

Serum activities of both aminotransferases vary with age in males and females [9,10]. Serum ALT activity hardly varies in boys aged from 4 to 18 years, whereas AST activity decreases significantly. Then, activities increase with a maximum at 45–55 years for ALT and 60 for AST, followed by a decrease for both the enzymes. In females, ALT and AST activities drop between 4 and 18 years, after which they increase progressively to a maximum at 65 years for ALT and 50 years for AST. Slight decreases are observed after these ages.

Variation of ALP activity with age is well known. During childhood and aldolescence, total serum ALP and the bone isoenzyme undergo similar changes. In boys the median is stable up to the age of 10. Thereafter there is a large increase in the dispersion up to 14 or 15 years, when the median is at a maximum. The decrease is then rapid and ALP reaches the value for adults at about 20 years (Fig. 2). Among girls the observed changes are similar but the maximum activity is reached at about 12 years and the adult value at about 17 to 18 years. ALP activity corresponding to the liver fraction, by contrast, is not much higher in young people than in adults [11,15].

Serum GGT activity in males increases with age between 18 and 55 years and afterwards decreases. In women, the activity increases after 18 years and maximal values are reached about 10 years later. In addition, the dispersion of the values increases with age [12].

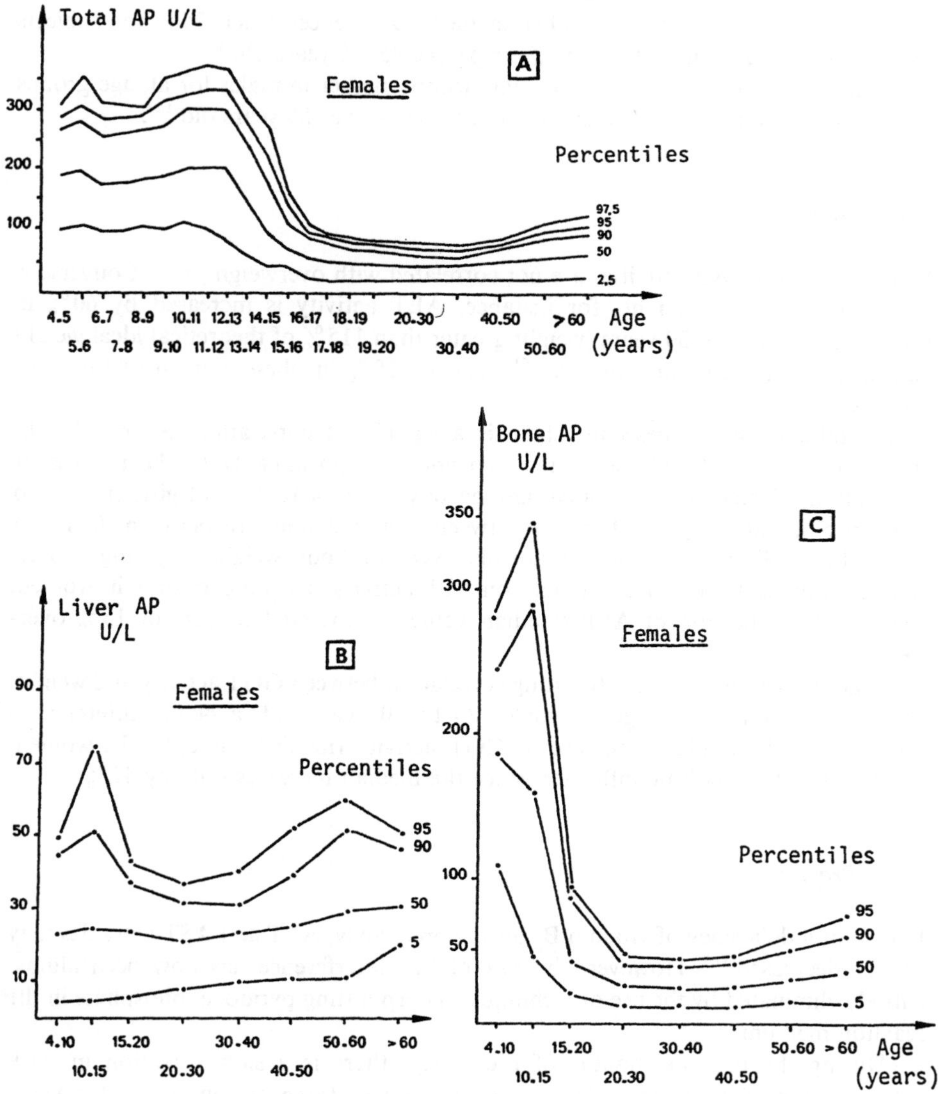

Fig. 2. Variation in the median and the dispersion of total (A), liver (B) and bone (C) alkaline phosphatase as a function of age in males [11, 15]

Serum AAP activity declines with age [13,14]. Children show higher AAP activities than adults. However, in adults the decrease with age is only significant for men [14].

2.3.2 Sex

Serum activities of both aminotransferases are higher in males than in females [9,10]. Total activity of ALP is 20% higher in men than in women; the liver isoenzyme does not depend on sex, whereas the bone isoenzyme is higher in men by about

40% [11]. GGT activity is also higher in men. Differences reach 25% for subjects 18–25 years old and more than 50% for people 45–55 years old [12].

Mean values for serum AAP are significantly higher in males for all age groups except for children under 10 years old and subjects over 55 years old [14].

2.3.3 Morphometry

Serum AST and AAP activities are not correlated with overweight [9, 14]. Conversely, ALT increases with weight; for instance, ALT activity is increased by 60% in obese men aged 20 to 30 years (weight greater than 115% of theoretical ideal weight calculated by the Lorentz formula [16]), and by 20% in obese women of the same group of age [10].

In children and adolescents, there is a significant correlation between height and both total ALP and the bone isoenzyme due presumably to the process of ossification. Correlation is positive among boys from 4 to 15 and girls from 4 to 10 years old, and is negative for boys between 15 and 20 and girls between 10 and 15 years old [11]. Similar correlations are observed for body weight in young people. In adults, the influence of overweight on ALP activity is significant only in women, where a 20% increase of ALP (median value) is observed in case of 15% overweight.

In adult men there is a fairly strong correlation between GGT activity and weight, apparently unrelated to age. In males 20 to 30 years old, a 40 kg difference is accompanied by a 145% increase in GGT activity (median value) [17]. In women, weight has relatively little influence, since the median increases only by 42%.

2.3.4 Pregnancy

In case of a deficiency of vitamin B_6 during pregnancy, ALT and AST activities may appear decreased [18]. However, the risk of this interference has now been almost entirely eliminated by the use of techniques incorporating pyridoxal phosphate in the reaction medium.

During the first six months of pregnancy, there is a slow elevation in ALP activity followed by a much larger increase (up to three or four times the usual values) during the last trimester, reaching a maximum at term. This increase is mainly due to the placental isoenzyme, whose activity may constitute 50% of the total activity [19, 20]. A recent report has shown that the bone isoenzyme is also increased in pregnancy [21].

Modifications of GGT in pregnancy are still a matter of debate, since some authors have found no variation of this enzyme during this period [22, 23] and others an increase [24] or a decrease [25].

When AAP is assayed with the substrate alanyl-4-nitroanilide, no difference in activity is found between pregnant and nonpregnant women. The increase in AAP in pregnant women when the enzyme activity is measured with the substrate leucyl-4-nitroanilide appears entirely due to the interference of another enzyme, cystine-aminopeptidase or oxytocinase, which is of placental origin [14].

2.3.5 Hormonal Status in Females

The influence of puberty and menopause on the five liver enzymes in serum is summarized in Table 2. Except GGT, all these enzymes seem to be affected by the hormonal status in females. Puberty is responsible for decrease in activities and menopause for increase.

Table 2. Influence of puberty and menopause on liver enzymes in serum

	Puberty	Menopause	References
ALT	↘5%	↗12%	10
AST	↘8%	↗11%	9
ALP	Total: ↘46% Bone: ↘53% Liver: no variation	Total: ↗25% Bone: ↗30% Liver: ↗22%	15, 26
GGT	No variation	No variation	12
AAP	↘5%	↗9%	14

Percentages correspond to variations of the median values

2.3.6 Blood Pressure

There are only few data concerning the effects of blood pressure on serum activities of liver enzymes. GGT seems to be elevated in hypertensive subjects [27,28]. Thus the median value is multiplied by three in subjects with a blood pressure higher than 150 mm Hg, in comparison with a reference population showing no sign of hypertension or pathological risk [28]. Variations of AAP depend on sex and are of limited size. They are observed only in subjects with a blood pressure higher than 160 mm Hg [14].

2.3.7 Genetic Factors

Serum AST and ALT activities have been found to be affected by genetic factors [29].

The influence of blood group on ALP activity is well known: Group 0 and B are associated with higher ALP activities than group A, while group AB is intermediate [30]. Race also affects serum ALP activity, which is greater in black children than in white children at puberty, owing to a greater development of the skeleton [20,31]. A recent study has shown differences in the thermal stability of the liver isoenzyme between black and white people [32]. Finally, it must also be kept in mind that the placental isoenzyme shows a genetic polymorphism, with three common alleles giving rise to six phenotypes, and numerous rare alleles; one of the placental genotypes may be associated with some cases of spontaneous abortion [33,34].

Genetic factors do not appear to influence the activities of the other liver enzymes.

2.3.8 Exercise

ALT and especially AST are increased by exercise, probably because of the release of muscle enzymes [35,36]. Elevations are slight after moderate exercise (2 to 6%). In contrast, a violent prolonged exercise may increase AST activity by 75% [37].

Slight increases due to exercise are also observed for ALP [35]. In the case of GGT or AAP, data are contradictory [38,39] or non-existent, respectively.

2.3.9 Rhythms

A circadian rhythm has been described for serum AST in which activity is greatest at 9:00 a.m. [40].

There is no circadian rhythm for ALP [41]. Conversely, a seasonal variation in ALP has been shown, with low values occurring in summer and high in winter; these changes could be due to changes in the bone isoenzyme [42].

No circadian or seasonal rhythms have been described for ALT, GGT and AAP.

2.3.10 Socio-professional Status

ALT, AST and GGT activities are higher in manual or agricultural workers than in sedentary professional men [9,10,12,17]. For GGT, the difference reaches 29% in men 20 to 30 years old and 50% in those aged 50 to 60 years [17].

2.3.11 Nutritional Diet

There are few data about the effects of nutritional diet on liver enzymes in serum. Reports essentially concern ALP, and are conflicting. Young [20] observed that a high carbohydrate-diet causes an increase in serum ALP in both men and women, while fasting and malnutrition reduce activity.

Conversely, Joseph and Kohn [43] observed no differences between malnourished children whose growth had been retarded and a control group of children with normal growth.

3 Drug Effects on Hepatic Enzymes in Serum

3.1 Origin of Drug Effects

A drug taken before the collection of a blood specimen can induce an increase or a decrease in the concentration of a plasma constituent. This effect can arise in two ways [44]. One is purely analytical: drug and/or its metabolites may at any stage affect the measurement of a constituent. In laboratory medicine this is called "analytical interference".

The second means is biological: the drug induces a change in a biological parameter by a physiological, pharmacological, or toxicological mechanism. This constitutes what is customarily called the effects of a drug on laboratory tests [45]. This very general term encompasses the unexpected secondary effects, desirable or undesirable, of drugs. It also sometimes includes the results of drug interactions, a term which has been proposed but which should be restricted to pharmacology.

When a toxicological mechanism is involved in the variations of liver enzymes in serum, variations are large (for instance, increase >3 times the upper reference limit for aminotransferases) and slight histological liver damage is observed. The liver lesions may be of different types (necrosis, hepatitis, cholestasis, steatosis, fibrosis). In addition, the effects are not necessarily dose- and duration-related [46]. Hepatotoxicity of many drugs has been demonstrated. This aspect of drug effects on hepatic enzymes will not be further developed here.

The other mechanisms correspond to harmless interactions, provoking often moderate variations with no histological liver damage. The effects are generally dose — and duration — dependent, but it is sometimes difficult to establish a clear-cut distinction between the different mechanisms responsible for a drug effect [46]. The physiological, pharmacological or biochemical mechanisms may involve induction or inhibition of enzyme synthesis, or modifications of its in vivo activity, or modifications of release due, for example, to variations of membrane structure [47].

3.2 Analytical Interference

Analytical interferences of drugs on aminotransferases, ALP and GGT were reviewed three years ago in a previous book [46, 48, 49].

Briefly, when AST is measured by the present optimized methods with pyridoxal phosphate, only ascorbic acid causes a modest elevation of the readings. When ALT is measured with such methods, no analytical interference has been described. Interferences were much more numerous with the colorimetric techniques, which are now obsolete (for review, see [46]).

Many drugs were reported to interfere with the measurement of ALP activity, depending on the method used. Amphotericin B, oxytetracycline, tetracycline, valproic acid, niflumic acid, dopexin, nitrofurantoin, ibuprofen, methotrexate, naproxen and methyldopa may produce apparent increases in ALP activity. Other drugs may cause decreases in measured activities: aminophylline and theophylline (Fig. 3) are the most representative examples of such an interference (review in [48]).

Four drugs were shown to interfere in the measurement of serum GGT, but at high concentrations: meprobamate, ascorbic acid, dopexin and propranolol [49, 50]. However, activities were modified by less than 6%. Until now, no information is available about drug interferences in the measurement of serum AAP activity.

3.3 Biological Effects

Biological effects of drugs on hepatic enzymes in serum were reviewed in previous publications [9–12, 46, 48, 49]. Therefore we only detail here the effects of three classes of

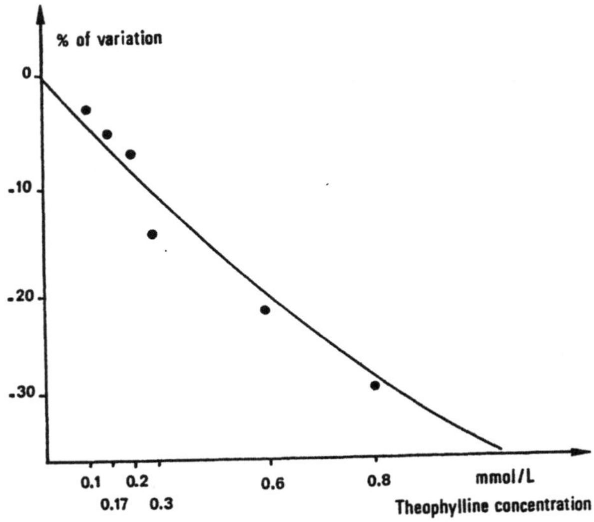

Fig. 3. Interference in vitro of theophylline in the method for measuring alkaline phosphatase activity [50]

drugs, which are currently administered and/or which have important effects on the enzymes.

3.3.1 Anticonvulsants

Most anticonvulsant drugs are enzyme inducers. Because they are frequently given in combination for a long time, these drugs cause numerous modifications and sometimes potentiate side effects. Moderate elevations of serum ALT (about 30–40%) are frequent in populations of epileptics receiving long-term treatment [51]. Phenytoin, carbamazepine, primidone and phenobarbital frequently cause hepatitis. Raised serum aminotransferase levels generally revert to baseline after the treatment is stopped [52–56]. Hepatotoxicity due to valproic acid is rare but has been reported [57,58]. In some cases, without clear hepatotoxocity, increases in serum aminotransferases have been observed. They seem to be related to the dose administered [59,60].

Hepatotoxicity of valproic acid, carbamazepine and phenytoin is responsible for increases in ALP activity [59,61,62]. Treatment with hydantoins or primidone accelerates the catabolism of vitamin D by induction, but in occasional individuals a decrease of the derivative 25-hydroxycholecalciferol is observed, resulting in osteomalacia and increased ALP activity [63].

Increased GGT activities after treatment with anticonvulsant drugs were observed as early as 1971 by Rosalki et al. [64]. Increases which may be related to enzyme induction, reach about 200% in adults and 400% in children [65]. They were described for drugs such as carbamazepine, primidone, phenobarbital and phenytoin given either alone or in combination [53,66,67].

The effect of anticonvulsants on AAP is not significant when studied in a whole treated population but increased activities are frequently observed in patients showing elevations of GGT due to induction by these drugs [14]. An additional hydrophilic isoform of AAP named F.AAP, seems to be often present in people consuming anticonvulsant drugs [68].

3.3.2 Hypolipidemic Agents

Long-term treatment with hypolipidemic agents causes moderate increases in serum aminotransferase activities [51,69,70]. Some cases of toxic hepatitis responsible for raised activities of aminotransferases have also been reported [71,72].

Hypolipidemic agents except colestipol decrease ALP activity up to 37% [51,73]. In adults, this change is due to the liver isoenzyme, and in children the liver and bone isoenzymes decrease simultaneously [74].

Lipid-lowering drugs such as clofibrate and fenofibrate generally decrease GGT activity up to 53% (mean value) [51,74–77]. Colestipol given alone seems to be without effect on GGT. Hypolipidemic agents decrease AAP activity in men, but in women, the variations are not significant [14].

3.3.3 Oral Contraceptives

The effects of oral contraceptives on aminotransferases are dependent on the estrogen dosages. A low estrogen level causes only a few abnormalities, whereas the older formulations containing higher estrogen levels give rise to increases of 20% to 40% of their activities [51,78].

The biological effects of oral contraceptives on serum ALP are controversial. Some authors have found a decrease of ALP activity in women taking these drugs [78–80]. For others, no change in ALP is induced by oral contraceptives [81,82].

Oral contraceptives are generally shown to be responsible for increases of GGT activity by 10% to more than 120% [79,83,88]. Elevations were reported to depend on the dose of estrogen administered [79] and on the duration of treatment [84]. Oral contraceptives also increase serum AAP activity and this change could be related to the presence of the F.AAP isoform [14,68].

3.3.4 Other Drugs

Antianginal drugs and anti-gout agents moderately increase ALT activity ; heparin is also responsible for elevation of both aminotransferases [9,10,46]. However, GGT activity is much more influenced by drugs than the other hepatic enzymes; the following groups of drugs are known to increase its serum activity [49]:

- Antidepressant agents
- Anticoagulants
- Anti-gout drugs
- Antianginal drugs
- Antiarrhythmic drugs
- Hypoglycemic agents
- Tranquilizers (to a lesser extent).

Besides the physiological or pharmacological effects described above, many drugs are known to modify the hepatic enzymes by hepatotoxic mechanisms and practically all drugs have been suspected of causing hepatic side effects in at least a few patients.

4 Effects of Alcohol and Tobacco

In a study of the effects of alcohol in a random population, Bagrel et al. [85] could find no significant alteration in serum ALT activity due to alcohol. Increases of ALT observed in alcoholics by other groups [86, 87] are probably related to hepatotoxicity of alcohol which is responsible for acute and chronic hepatitis.

Conversely, the consumption of alcohol produces an increase in serum AST activity [87, 88]. Recently mitochondrial AST (mAST) and especially mAST/total AST ratio were proposed as sensitive biological markers to discriminate liver diseases for alcoholic and non alcoholic patients [89, 90]. However, increases of mAST due to alcohol intake are not significant in an unselected population [91] and the use of mAST as a marker of excessive alcohol consumption has not been, until now, really validated. Total AST activity seems to be slightly lower in current smokers than in non-smokers [88].

Moderate consumption of alcohol produces no significant change in serum ALP activity, which therefore is presumably not inducible by alcohol in contrast with GGT [86]. ALP was shown to be higher among current smokers than nonsmokers and exsmokers [88].

Much has been reported about the effect of alcohol intake on the measurement of serum GGT activity. In 1970, Rosalki's group [92] showed for the first time the increase in GGT activity induced by alcohol. This increase is found in 50% to 75% of excessive drinkers. GGT activity may be multiplied by 20 in alcoholics [49, 93]. The measurement of this enzyme is a valuable test for detecting alcohol consumption and also for monitoring detoxication of alcoholics [94].

Serum AAP activity is significantly increased in subjects drinking more than 44 g of alcohol a day [14] and, in these people, the F-AAP isoform is present at a high frequency [68]. In addition, two recent studies have shown that smoking increases serum AAP activity [13, 14].

5 Conclusion

The study of the biological factors of variation of the hepatic enzymes is necessary to establish reference values allowing a correct interpretation of the measured activities.

Besides the well-known hepatotoxicity of many drugs, which is responsible for large variations of the enzymes in serum, it is also necessary to take into account the modifications of activity due to physiological or pharmacological effects of drugs, such as induction. These effects are generally characterized by a moderate variation without histological liver damage, but they occur relatively frequently in the population.

In other words, the observed variations due to drug intake may also be used to evaluate the effects of drugs on the liver tissue and its enzymes. This is especially of interest in the biochemical and clinical trials necessary for the development of new drugs. For this purpose and especially to assess enzyme induction, in addition to the classical measurement of GGT activity, clinical chemists resort to the determination of various compounds: enzymes of circulating blood cells, endogenous metabolites such as 6-beta-hydroxycortisol and glucaric acid, drugs such as aminopyrine and antipyrine and their metabolites; all these compounds are indirect markers of the drug effects on the liver enzymes.

6 References

1. de Broe ME, Roels F, Nouwen EJ, Claeys L, Wieme RJ (1985) Hepatology 5: 118
2. Rej R (1985) CRC Crit Rev Clin Lab Sci 21: 99
3. Moss DW (1987) Clin Biochem 20: 225
4. Artur Y, Wellman-Bednawska M, Siest G (1984) Ann Biol Clin 42: 115
5. Sanderink GJ, Artur Y, Galteau MM, Wellman-Bednawska M, Siest G (1986) Electrophoresis 7: 471
6. Moss DW (1986) Clin Chim Acta 161: 123
7. Siest G (1985) in: Siest G, Henny J, Schiele F, Young DS (eds) Interpretation of clinical laboratory tests. Reference values and their biological variation. Biomedical Publ, Foster City, p. 3
8. Siest G (1981) in: Grasbeck R, Alstrom T (eds) Reference values in laboratory medicine. The current state of art. Wiley and Sons, New York, p 45
9. Vincent-Viry M, Galteau MM, Schiele F (1985) in: Siest G, Henny J, Schiele F, Young DS (eds) Interpretation of clinical laboratory test. Reference values and their biological variation. Biomedical Publ, Foster City, p 130
10. Vincent-Viry M, Galteau MM, Schiele F (1985) in: Siest G, Henny J, Schiele F, Young DS (eds) Interpretation of clinical laboratory tests. Reference values and their biological variation. Biomedical Publ, Foster City, p 69
11. Henny J, Schiele F (1985) in: Siest G, Henny J, Schiele F, Young DS (eds) Interpretation of clinical laboratory tests. Reference values and their biological variation. Biomedical Publ, Foster City, p 96
12. Siest G, Schiele F, Artur Y (1985) in: Siest G, Henny J, Schiele F, Young DS (eds) Interpretation of clinical laboratory tests. Reference values and their biological variation. Biomedical Publ, Foster City, p 235
13. Mueller PW, Phillips DL, Steinberg KK (1987) Clin Chem 33: 363
14. Sanderink GJ, Artur Y, Schiele F, Gueguen R, Siest G (1988) Clin Chem 34: 1422
15. Schiele F, Henny J, Hitz J, Petitclerc C, Gueguen R, Siest G (1983) Clin Chem 29: 4
16. Locuty J, Kuntz C, Guillemot M (1985) in: Siest G, Henny J, Schiele F, Young DS (eds) Interpretation of clinical laboratory tests. Reference values and their biological variation. Biomedical Publ, Foster City, p 33
17. Schiele F, Guilmin AM, Detienne H, Siest G (1977) Clin Chem 23: 1023
18. Hafkenscheid JCM, Dijt CCM (1979) Clin Chem 25: 55
19. Katterman R (1976) Deutsch Med Wochenschr 101: 1508
20. Young DS (1979) in: Brown SS, Mitchel FL, Young DS (eds). Clinical diagnosis of diseases. Elsevier, Amsterdam, p 1
21. Valenzuela GJ, Munson LA, Tarbaux NM, Farley JR (1987) Clin Chem 33: 1801
22. Walker FB, Hoblit DL, Cunningham FG, Combes B (1974) Obstet Gynecol 43: 745

23. Herzberg M, Tenenbaum E, Fischel B, Wiener MH (1977) Clin Chem 23: 596
24. Jones DD, Williams G, Prochazka B (1974) Enzyme 17: 139
25. Feldman HU, Pfeiffer R, Hirche H (1974) Deutsch Med Wochenschr 99: 1171
26. Crilly RG, Jones MM, Horsman A, Nordin BEC (1980) Clin Sci 58: 341
27. Henningsen NC, Ohlsson O, Mattiasson I, Trell E, Kristenson H, Hood B (1980) Acta Med Scand 207: 245
28. Hitz J, Steinmetz J, Henny J, Siest G. (1984) Ann Biol Clin 42: 289
29. Whitfield JB, Martin NG (1985) Enzyme 33: 61
30. Tschanz C, Dauwalder H, Colombo JP (1974) Clin Biochem 7: 68
31. Bennett DL, Ward MS, Daniel WA (1976) J Pediatr 88: 633
32. Groff J, Rutkowski R (1988) Clin Chim Acta 171: 119
33. Beckman G, Beckman L (1984) in: Stigbrand T, Fishman WH (eds) Human alkaline phosphatases. AR Liss, New York, p 15
34. Sanderink GJ, Artur Y, Paille F, Galteau MM, Siest G (1988) Clin Chem 34: 730
35. Galteau MM, Siest G (1973) in: Siest G (ed) Reference values in human chemistry. Karger, Basel, p 221
36. Statland BE, Winkel P, Bokelund H (1975) in: Siest G (ed) Organisation des laboratoires, Biologie Prospective. Expansion Sci Française, Paris, p 717
37. Siest G, Galteau MM (1974) Enzyme 17: 179
38. Friedel R, Mattenheimer H, Trautschol DI, Forster G (1976) J Clin Chem Clin Biochem 14: 109
39. Haralambie G (1976) Clin Chim Acta 72: 363
40. Halberg F (1977) Hosp Pract 12: 139
41. Statland BE, Winkel P, Bokelund H (1973) Clin Chem 19: 1373
42. Devgun MS, Paterson CR, Martin BT (1981) Enzyme 26: 301
43. Joseph MV, Kohn TK (1978) Ind Pediatr 15: 45
44. Siest G, Galteau MM, Notter D et les Membres de la Commission "Effets des Médicaments sur les Examens de Laboratoire" de la Société Française de Biologie Clinique (1981) Ann Biol Clin 39: 91
45. Siest G, Dawkins SJ and the members of the IFCC EP "Drug Effects in Clinical Chemistry" (1984) J Clin Chem Clin Biochem 22: 271
46. Vincent-Viry M, Delwaide P (1985) in: Siest G, Galteau MM, Schiele F, Henny J (eds) Examens de laboratoire et médicaments. Interférences analytiques et variations pharmacologiques. Expansion Sci Française, Paris, p 259
47. Siest G (1985) in: Siest G, Galteau MM, Schiele F, Henny J (eds) Examens de laboratoire et médicaments. Interférences analytiques et variations pharmacologiques. Expansion Sci Française, Paris, p 1
48. Vincent-Viry M, Gouy D (1985) in: Siest G, Galteau MM, Schiele F, Henny J (eds) Examens de laboratoire et médicaments. Interférences analytique et variations pharmacologiques. Expansion Sci Française, Paris, p 181
49. Artur Y, Gouy D (1985) in: Siest G, Galteau MM, Schiele F, Henny J (eds) Examens de laboratoire et médicaments. Interférences analytiques et variations pharmacologiques. Expansion Sci Française, Paris, p 110
50. Gosset J (1983) Etude des interférences analytiques et des effets pharmacologiques des médicaments sur les examens de laboratoire, Thèse 3ème cycle, Biochimie Pharmacologique, Nancy
51. Siest G, Galteau MM, Schiele F (1980) In: Siest G, Young DS (eds) Drug measurement and drug effects in laboratory health science. Karger, Basel, 1980, p 204
52. Aiges HW, Daum F, Olson M, Kahn E, Teichberg S (1980) J Pediatr 97: 22
53. Gauchel FD, Lehr HJ, Gauchel G, Von Harnack GA (1973) Dtsch Med Wochenschr 98: 1391
54. Hopen G, Nesthus L, Laerum OD (1981) Acta Med Scand 210:333
55. Josephs SH, Rohtman SJ, Buckley RH (1980) J. Allergy Clin Immunol 66: 166
56. Rigau-Canardo J, Brugera M, Sese J, Morlans G (1984) Gastroenterol Clin Biol 8: 769
57. Ware S, Millward-Sadler CH (1980) Lancet 2: 110
58. Zafrani EJ, Berthelot P (1982) Hepatology 2: 648
59. Sussman NM, Mac Lain LW (1979) JAMA 242: 1173
60. Willmore LJ, Wilder BJ, Bruni J, Villarreal HJ (1978) Neurology 28: 961
61. Grunnar H, Ingerid N, De Didrik L (1981) Acta Med Scand 210: 333
62. Mullick FG, Ishak KG (1980) Am J Clin Pathol 74: 442

63. Pylypchuk G, Oreopoulos DG, Wilson DR, Harrison JE, Mc Neill KG, Meema HE, Ogilvi ER, Sturtridge WC, Murray TM (1978) Can Med Assoc J 118: 635
64. Rosalki SB, Tarlow D, Rau D (1971) Lancet 2376
65. Batt AM, Siest G, Khodjet El Khil R, Le Perron B, Weber M, Tridon P (1976) in: Siest G, Young DS (eds) Drug interference and drug measurement in clinical chemistry. Karger, Basel, p 33
66. Heipertz R, Eickhoff K, Poser W (1978) Klin Wochenschr 56: 921
67. Ohnhaus EE, Park BK (1979) Eur J Clin Pharmacol 15: 139
68. Sanderink GJ, Artur Y, Paille F, Siest G (1989) Clin Chim Acta 179: 23
69. De Gennes JL, Truffert J, Periac P (1978) Nouv Presse Med 7: 2398
70. Delwaide PA, Jadin A, Heusghem C (1973) in: Heusghem C, Lechat P (eds) Les effets indésirables des médicaments. Masson, Paris, p 698
71. Cloarec M (1978) Vie Med 31: 2781
72. Vachon JM (1980) Nouv Presse Med 9: 2740
73. Schade RWB, Demacker PNM, Van't Laar A (1975) Lancet 1: 862
74. Steinmetz J, Gaspart E, Notter D (1980) in: Siest G (ed) Drug effects on laboratory test results. Martinus Nijhoff, The Hague, p 295
75. Ferrari C, Testori GP, Scanni A, Frezzati S, Bertazzoni A, Romussi M (1976) N Engl J Med 295: 449
76. Schiele F, Neuman JL, Le Perron B, Galteau MM, Siest G (1980) in: Siest G, Young D (eds) Drug measurement and drug effects in laboratory health science. Karger, Basel, p 185
77. Probstfield JL, Statland BE, Gorman L, Hunninghake DB (1983) Metabolism 32: 818
78. Batt AM, Siest G, Loppinet V, Guerin P, Gueguen R, Floc'h AY (1974) Ann Biol Clin 32: 245
79. Herbeth B, Bagrel A, Dalo B, Siest G, Leclerc J, Rauber G (1981) Clin Chim Acta 112: 293
80. Moore B, Paterson MEL, Sturdee DW, Whitehead TP (1981) Br J Obstet Gynaecol 88: 853
81. Dickerson J, Bressler R, Christian CD (1980) Contraception 22: 597
82. Miale JB, Kent JW (1974) Am J Obstet Gynecol 120: 264
83. Siest G, Batt AM, Galteau MM, Weber M, Tridon P (1974) Thérapie 29: 907
84. Straus B, Calic-Runje R, Cepelak I (1982) Acta Pharm Jugosl 32: 191
85. Bagrel A, D'Houtaud A, Gueguen R, Siest G (1979) Clin Chem 25: 1242
86. Freer DE, Statland BE (1977) Clin Chem 23: 830
87. Patels S, O'Gorman P (1975) J Clin Pathol 28: 414
88. Chan-Yeung M, Ferreira P, Frohlich J, Schulzer M, Tan F (1981) Am J Clin Pathol 75: 320
89. Panteghini M, Malchiodi D, Calarco M, Bonora R (1984) J Clin Chem Clin Biochem 22: 153
90. Nalpas B, Vassault A, Charpin S, Lacour B, Berthelot P (1986) Hepatology 6: 608
91. Schiele F, Artur Y, Varasteh M, Wellman M, Junge J, Siest G (1987) Enzyme 38: Suppl 1, 25 (abstract)
92. Rosalki SB, Rau D, Lehman D, Prentice M (1970) Ann Clin Biochem 7: 143
93. Salaspuro M (1987) Enzyme 37: 87
94. Lamy J, Baglin MC, Ferrant JP, Weill J (1975) Clin Chim Acta 60: 103

Plasma Amylase in Pancreatic and Hepatobiliary Disease

Sidney B. Rosalki

Department of Chemical Pathology and Human Metabolism, Royal Free Hospital and School of Medicine, London, UK

Plasma amylase measurement is of established value in the investigation of pancreatic disease. It is the most popular diagnostic procedure for the confirmation of suspected acute pancreatitis, combining high clinical sensitivity with analytical reliability and convenience. Dry chemistry reagents (test strips) can give rapid, accurate quantification of amylase activity in either plasma or whole blood to give immediate test results.

The sensitivity of plasma amylase determination is increased by isoenzyme (isoamylase) examination. Measurement of the specific pancreatic (P) fraction is easily carried out after inhibition of salivary (S) amylase by wheat protein or antibody, or immunoprecipitation of the salivary isoenzyme. Electrophoresis can provide an even more complete picture of the isoenzyme composition of plasma — the P3 fraction having high specificity for acute pancreatitis and identifying complications.

Low plasma P-isoamylase activity serves to identify pancreatic exocrine insufficiency in chronic pancreatitis and cystic fibrosis and provides a simple non-invasive procedure of comparable sensitivity to other pancreatic enzyme and non-enzyme tests of pancreatic exocrine function.

Liver tissue contains little amylase activity. Small amounts of amylase are detectable in bile, suggesting a possible excretory mechanism. In hepatobiliary tract disease, plasma amylase examination identifies pancreatic involvement accompanying alcoholic liver disease, primary biliary cirrhosis and primary sclerosing cholangitis. In the former, amylase changes are largely independent of liver histology and frequently may be rapidly reversed with abstinence. Amylase changes in plasma encountered in primary biliary cirrhosis and primary sclerosing cholangitis suggest that the duct changes present in the liver may be mirrored in the pancreas.

Progress in Clinical Biochemistry
and Medicine, Vol. 8

1 Introduction

Whilst changes in plasma amylase activity in pancreatic disease are well recognised, altered plasma amylase in liver and hepato-biliary disease is less familiar. Here, changes in plasma amylase in acute pancreatitis, chronic pancreatitis and cystic fibrosis will be considered, and abnormalities encountered in liver disease, especially alcoholic liver disease, primary biliary cirrhosis and primary sclerosing cholangitis described.

2 Plasma Amylase in Pancreatic Disease

2.1 Acute Pancreatitis

This disorder accounts for 0.5 percent of general hospital admissions and 5–10 percent of hospital admissions for acute abdominal pain [1,2]. It is most frequently a complication of biliary tract disease, particularly gallstones, and of alcoholism, and is characterised clinically by severe abdominal or back pain, shock, nausea, vomiting and ileus. Resolution is usually complete, but complications include pancreatic necrosis, abscess, or pseudocyst formation, found in 2–5 percent of patients. The disorder carries a 10 percent mortality. Operation is generally inadvisable in acute pancreatitis, so it is essential to differentiate it from other acute abdominal disorders which require surgery.

2.1.1 Plasma Amylase

Amylase is the plasma enzyme most frequently measured for the confirmation of suspected acute pancreatitis, combining high clinical sensitivity with analytical reliability and convenience. An added advantage is that amylase activity can be rapidly and accurately measured in the Emergency Clinic or near the bedside on whole blood or plasma by the use of dry chemistry reagent strip methods. The enzyme exceeds the upper reference limit in 90 percent of patients, at which level specificity approximates to 80 percent [3]. Plasma amylase starts to rise approximately 4 hours after onset of pain, peaks at 24 hours and remains elevated from 3 to 7 days, but with more prolonged elevation with pseudocyst formation. The plasma amylase level shows little relation to the severity of the pancreatitis. A raised plasma amylase is not specific for pancreatitis and false positive elevation may result from many other causes, especially other acute abdominal disorders, and from impaired excretion of the enzyme in renal disease [4]. The higher the activity, the greater the likelihood of pancreatitis. At a cut-off level of twice the upper reference limit, specificity is increased by some 10 percent, with little loss of sensitivity [5]. At five times the upper reference limit, specificity increases to almost 95 percent but sensitivity declines to 60 percent.

In the interpretation of increased amylase values, it is important to recognise that ethnic differences exist in plasma amylase activity in normal subjects [6]. As a result, in the United Kingdom, one half of West Indians and up to one third of Indians show amylase values above the "usual" upper reference limit, with values respectively averaging twice and one and a half times those of indigenous Caucasians. In consequence, their normal total amylase might be misinterpreted as abnormal and indicative of pancreatic disease if their ethnic origin is ignored.

2.1.2 Plasma Isoamylase

One way which the plasma amylase determination for pancreatic disease may be increased is by plasma isoamylase measurement. Plasma amylases can be separated into those of pancreatic (P) and salivary (S) types. Pancreatic, abdominal and biliary tract disease result in elevation usually of P-type amylase. Salivary disease, pregnancy disorders, lung disease and diabetes give rise to plasma amylase elevation which is usually of S-type. Renal failure, diabetic ketoacidosis, and macroamylasemia may elevate either the P or S fractions [7]. When hyperamylasemia accompanies abdominal pain, diagnostic accuracy for pancreatic disease is improved by some 20 percent by isoamylase measurement [8].

The P- and S-isoamylases may be rapidly and conveniently measured by the inclusion in the reagent of a wheat-derived inhibitor of salivary amylase [9]. Inhibited enzyme then represents the S-fraction and residual activity P-fraction. More recently, immunoprecipitation has been used to remove [10, 11], or immunoinhibition to inhibit [12], S-amylase, followed by measurement of unprecipitated or uninhibited P-enzyme. Isoamylases may also be separated by electrophoresis [13, 14] or by chromatography [15] or measured by radioimmunoassay [16], but these are more time-consuming than the inhibition assays. P-isoamylase determination increases diagnostic sensitivity and specificity for acute pancreatitis by some 10 percent compared with total amylase measurement, and elevation is more prolonged by 1 or 2 days.

Though less convenient than inhibition methods, electrophoretic demonstration and quantification of amylase isoenzymes may be especially valuable for pancreatitis confirmation, since a specific amylase isoenzyme demonstrable only by electrophoresis appears in plasma in acute pancreatitis. Electrophoresis of normal plasma demonstrates isoamylases of both pancreatic and salivary origin, usually the P2 and S1 fractions. In acute pancreatitis, an isoamylase of electrophoretic mobility intermediate between these isoenzymes and known as P3 isoamylase is found [17]. In suspected acute pancreatitis, plasma P3 isoamylase demonstration has a diagnostic sensitivity, specificity and accuracy approaching 100 percent [3, 18]. However, because of laboratory delay inevitable in carrying out full isoamylase analysis, the procedure is generally restricted to difficult diagnostic problems.

Plasma P3 isoamylase elevation persists for some 2–3 days following return of total amylase to normal, the duration of elevation exceeding that of P-isoamylase by inhibition assay [19]. More persistant elevation in excess of 14 days carries an unfavourable prognosis [20]. P3 isoamylase in excess of 15 percent of total amylase then suggests pseudocyst formation with a 90 percent sensitivity and specificity [21]. The P3 isoamylase is thought to arise from post-translational modification of P2 isoamylase as a consequence of removal of terminal glutamine and asparagine from the pancreatic isoenzyme by protease, the resulting increased negative charge giving rise to a more anodal (so-called "aged" amylase) P3 form [22,23]. In acute pancreatitis, protease activation within the pancreas and release into the plasma converts the usual P2 form into the P3 fraction. In pseudocyst, isoamylase sequestered within the cyst is converted to the P3 form, which is gradually transported across the cyst membrane to give plasma P3 elevation.

2.2 Chronic Pancreatitis

In contrast to the increased plasma enzyme activity in acute pancreatitis, diminished activity may be useful for the detection of impaired pancreatic function from chronic pancreatitis. This chronic inflammatory disorder of the pancreas accounts for 20 percent of patients presenting with non-specific chronic abdominal symptoms [1]. It is due principally to alcoholism. It is characterised by recurrent pain in the abdomen or back, together with evidence of pancreatic exocrine insufficiency from gland atrophy and resulting in steatorrhoea and malabsorption. Diabetes may also develop.

2.2.1 Amylase and Isoamylase

Although total amylase in plasma remains within the normal range, the pancreatic isoamylase becomes markedly reduced in pancreatic insufficiency and its measurement provides an inexpensive, non-invasive screening procedure which has been shown to correlate well with other tests of pancreatic exocrine function including duodenal juice enzyme measurement [24–26]. Reduced P-isoamylase is best identified by electrophoresis with densitometric quantification, but may also be conveniently quantified after wheat-protein or immuno-inhibition or immunoprecipitation of S-amylase.

Pancreatic isoamylase in plasma is reduced in 70 percent of patients with chronic pancreatitis overall, this frequency being increased to 85 percent in advanced disease and reducted to 50 percent in less severe insufficiency [27,28]. When chronic pancreatitis is suspected, a low pancreatic isoamylase indicates reduced pancreatic exocrine function with a positive predictive value of 95 percent, whereas normal pancreatic isoamylase excludes severe insufficiency with 90 percent probability [26]. Pancreatic isoamylase measurement may also be used to demonstrate impaired exocrine function in many diseases other than chronic pancreatitis, including cystic fibrosis [29] and diabetes [30].

2.3 Cystic Fibrosis

Cystic fibrosis is a disorder of mucous exocrine glands which occurs with a United Kingdom frequency of approximately 1 per 2000 live births, and is inherited as an autosomal recessive trait. One in 20 adults are heterozygote carriers of the gene for the disorder. In children with the disorder, pancreatic duct obstruction by thick mucous is accompanied or followed by gland atrophy with pancreatic insufficiency in more than 85 percent of patients.

2.3.1 Isoamylases

Pancreatic insufficiency in cystic fibrosis is conveniently confirmed by plasma isoamylase measurement [29,31]. Total amylase activity generally remains within the reference range, due to normal or increased enzyme activity of salivary origin. Pancreatic isoamylase activity in plasma is, however, generally absent or grossly diminished. Eighty to 90 percent of children or young adults show reduced values. It is essential to compare values with those of an aged-matched reference population, for pancreatic amylase is low or absent in the pancreas and plasma of the neonate [32]. P-isoamylase only starts to be useful after the age of 18 months [33]. Pancreatic isoamylase measurement in cystic fibrosis reflects both the expression and the progression of the pancreatic exocrine insufficiency usually present. It correlates well with other tests of exocrine function, including duodenal juice enzyme measurements. Activity is unaffected by oral pancreatic enzyme replacement therapy and can be used as an indication for such therapy, which is unnecessary if plasma P-isoamylase activity is normal.

3 Plasma Amylase in Hepatobiliary Disease

3.1 Amylase in Liver and Bile

The liver contains very little amylase activity, less than 100 units per kilogram of homogenised tissue [34], equivalent to an activity less than 100 units per litre, so that the liver to serum amylase ratio is generally less than one [35]. It may, therefore, be assumed that the amylase content of the liver principally derives not from the hepatocyte, but from the blood reaching the liver tissue. Isoamylase examination reveals the amylase of liver to be of S- and P-type, especially the former.

Bile also contains low amylase activity, again predominantly S and again, equal to or lower than the corresponding serum value [36,37]. In animals, some hepatobiliary excretion of systemically injected P-amylase has been observed, but the amount excreted is less than 5 percent of the injected dose [38]. Salivary isoamylase in plasma is glycosylated and is thought to be taken up by glycoprotein receptors on the hepatocyte surface, following which it may pass through the hepatocyte to reach the bile through fusion of endocytic vesicles with the membrane of bile canaliculi [39,41]. That this excretory mechanism may also operate in man is suggested by the increase in S-amylase activity in plasma which follows partial hepatectomy [39]. It is likely, therefore, that impaired hepatobiliary excretion of plasma amylase may contribute to the increase of S and P isoamylases found in hepatobiliary disease despite the absence of a true liver amylase. However, the main contribution to plasma amylase changes in hepatobiliary disease is believed to be concomitant pancreatic and salivary disease.

Three hepatobiliary disorders will be especially considered, alcoholic liver disease, primary biliary cirrhosis and primary sclerosing cholangitis.

3.2 Alcoholic Liver Disease

Alcohol is now responsible for more than 10 percent of acute pancreatitis and more than 50 percent of chronic pancreatitis in the United Kingdom. In addition, many alcoholics show salivary gland hyperplasia. It is likely, therefore, that amylase and isoamylase changes observed in alcoholics result from involvement of these organs.

3.2.1 Amylase and Isoamylase

Overall, the frequency of plasma amylase change increases with the severity of alcoholism. Total amylase, P- and S-isoamylase increase in found in some 10 percent of acutely intoxicated patients both alcoholics and non-alcoholics [41]. In alcoholics hospitalised for detoxication or rehabilitation, the frequency of amylase and isoamylase elevation increases to 15 percent [42]. Approximately 20 percent of alcoholics admitted to hospital with alcoholic liver disease, but without clinical pancreatic disease show increased total amylase activity, with a similar percentage showing increased pancreatic isoamylase. The salivary isoenzyme is also increased in 20 percent [43]. All increases are generally modest and usually below twice the upper reference limit. There is no significant relationship between increased plasma amylase activity or the frequency of enzyme change and the liver histology, though patients with alcoholic liver cirrhosis show the highest mean values: perhaps reflecting the long duration of their alcoholism and an increased likelihood of pancreatic injury.

Approximately 15 percent of patients with alcoholic liver disease but without obvious pancreatic disorder show decreased pancreatic isoamylase, suggesting sub-clinical pancreatic exocrine insufficiency [43]. The frequency of subnormal pancreatic isoamylase increases with the duration of alcoholism [42]. Pancreatic exocrine insufficiency in the alcoholic may, however, be encountered as an acute effect rather than an indication of chronic pancreatitis, since it may be rapidly reversible with abstinence [44]. Temporary pancreatic insufficiency from alcoholism may contribute to the gastro-intestinal disturbance and malabsorption known to follow drinking bouts.

3.3 Primary Biliary Cirrhosis

Primary biliary cirrhosis is an autoimmune disease of the liver predominantly of females, which presents with jaundice, itching and hepatomegaly. Sjogren's syndrome with salivary gland hyposecretion is a frequent accompaniment. The disorder causes destruction and obstruction of the small intrahepatic ducts and periductular inflammation and proceeds to fibrosis. Similar duct changes may be found in the pancreas and in the salivary glands.

3.3.1 Amylase and Isoamylase

In primary biliary cirrhosis, plasma total amylase and pancreatic isoamylase are increased in about one third of subjects, salivary isoamylase is increased in 20 percent [45,46]. These changes suggest that in these patients, the liver disorder is accompanied by pancreatic and/or salivary gland pathology. The plasma enzyme increase may reflect the presence of inflammatory changes in these glands or duct destruction or obstruction with reflux of enzyme into the plasma. Reduced P-isoamylase in plasma is uncommon (approximately 5 percent).

3.4 Primary Sclerosing Cholangitis

Primary sclerosing cholangitis is a cholestatic liver disorder thought to be due to portal toxaemia, it frequently accompanies ulcerative colitis. It is more frequent in males and presents with fever, jaundice and itching. Histologically, the interlobular ducts and surrounding tissue show a chronic inflammatory mononuclear infiltration and the disorder proceeds to fibrosis. Similar changes may be found in the large extrahepatic and pancreatic ducts.

3.4.1 Amylase and Isoamylase

Approximately 40 percent of patients show increased plasma total amylase and pancreatic isoamylase [45,46]. Salivary isoamylase is only occasionally raised (approximately 15 percent), reflecting the infrequency of salivary involvement in this disorder. A decrease in pancreatic amylase is observed in approximately 20 percent of patients, presumably from gland atrophy secondary to inflammatory change.

3.5 Comparison of Plasma Amylase in Primary Biliary Cirrhosis and Primary Sclerosing Cholangitis

If changes in primary biliary cirrhosis and primary sclerosing cholangitis are compared, the frequency of abnormality of total amylase and plasma pancreatic amylase is similar [46], though an early report suggested a higher incidence in the former [45]. Salivary amylase is more frequently raised in primary biliary cirrhosis and P-amylase more often decreased in primary sclerosing cholangitis, though the differences do not reach statistical significance. The plasma P- and S-amylase changes in these disorders most likely reflect the known high frequency of pancreatic and salivary involvement. They do not appear to reflect impaired hepatic excretion of plasma enzymes since values do not correlate with other indicators of cholestasis such as elevation of bilirubin or alkaline phosphatase [45]. Nevertheless, some contribution from impaired hepatic uptake cannot be discounted because of the high frequency of plasma S-isoamylase increase encountered in primary biliary cirrhosis, a disorder in which *reduced* S-isoamylase might be anticipated from the high frequency (up to 70 percent) of accompanying Sjogren's syndrome with salivary *hypo*secretion in this disease.

3.6 *Mechanism of Plasma Amylase Changs in Liver Disorders*

In alcohol liver disease, primary biliary cirrhosis and primary sclerosing cholangitis, it may be concluded that the P-isoamylase increase reflects pancreatic involvement since other pancreatic enzymes are known to increase also, and each disorder may be accompanied by pancreatic histological abnormality. That P-isoamylase decrease reflects accompanying exocrine insufficiency, is supported by intubation studies that show that duodenal enzyme output is reduced [45]. S-amylase increase suggests accompanying salivary gland damage, all three liver disorders being known to show clinical or histological salivary gland changes. In none of the liver disorders studied is the liver itself thought to be a source of plasma amylase, though, particularly with S-amylase increase, a contribution from diminished hepatic uptake or excretion of plasma amylase cannot be excluded.

3.7 *Other Liver Disorders*

Increases of plasma amylase have been observed with low frequency in many other varieties of liver disease. In acute viral hepatitis, 15 percent of patients may show slight plasma elevation, usually of P-type [35,47]. In non-alcoholic cirrhosis, the frequency of total amylase increase approximates to 20 percent with P-isoamylase elevation in 30 percent [48]. Chronic active hepatitis rarely shows plasma amylase elevation [49], increased plasma amylase has however been observed in some 10 percent of such patients, with 20 percent showing increased P-isoamylase and 10 percent increased S [50]. In all these varieties of liver disease, the degree of elevation is slight and less than twice the upper reference limit.

4 Concluding Remarks

The value of plasma amylase determination in the diagnosis of acute pancreatic disease has stood the test of time. Rapid and specific diagnosis is facilitated by recent methodological improvement and isoamylase measurement and the latter conveniently permits the identification of pancreatic exocrine insufficiency. The plasma amylase and isoamylase changes encountered in hepatobiliary disorders are thought to reflect concomitant pancreatic and salivary gland involvement, though a contribution (particularly to plasma S-isoamylase) from impaired hepatic uptake or biliary excretion of the plasma amylase cannot be excluded.

5 References

1. Braganza J (1982) Does your patient have pancreatic disease? J R Coll Phys, Lond 16: 13–22
2. Dombal FT de (1979) Acute abdominal pain. An OMGE survey. Scand J Gastroenterol 14 (suppl 1): 29–43

3. Massey TH (1985) Efficiency in the diagnosis of acute pancreatitis increased by improved electrophoresis of amylase isoenzyme P3 on cellulose acetate. Clin Chem 31: 70–75
4. Salt W, Schenker S (1976) Amylase, its clinical significance: a review of the literature. Medicine 55: 269–283
5. Steinberg WM, Goldstein SS, Davies ND, Shamma'a J, Anderson K (1985) Diagnostic assays in acute pancreatitis. A study of sensitivity and specificity. Ann Intern Med 112: 576–580
6. Tsianos EB, Jalali MT, Gowenlock AH, Braganza JM (1982) Ethnic hyperamylasaemia: Clarification by isoamylase analysis. Clin Chim Acta 124: 13–21
7. Berk JE, Fridhandler L (1981) Hyperamylasaemia: interpretation and newer approaches to evaluation. Adv Internal Med 26: 235–264
8. Koehler DF, Eckfield JH, Levitte MD (1982) Diagnostic value of routine isoamylase assay of hyperamylasaemia serum. Goestroenterol 82: 887–890
9. O'Donnell MD, Fitzgerald O, McGeeney KF (1977) Differential serum amylase determination by use of an inhibitor and design of a routine procedure. Clin Chem 23: 560–566
10. Mifflin TE, Benjamin DC, Bruns DE (1985) Rapid quantitative specific measurement of pancreatic amylase in serum with use of a monoclonal antibody. Clin Chem 31: 1283–1288
11. Gerber M, Naujoks K, Lenz H, Gerhardt W, Wulff K (1985) Specific immunoassay of amylase isoenzymes in human serum. Clin Chem 31: 1331–1334
12. Gerber M, Naujoks K, Lenz H, Wulff K (1987) A monoclonal antibody that specifically inhibits human salivary α-amylase. Clin Chem 33: 1158–1162
13. Meites S, Rogols S (1971) Amylase isoenzymes. CRC Crit Rev Clin Lab Sci 2: 103–138
14. Foo Y, Rosalki SB (1984) Isoenzymes of α-amylases. In: Bergmeyer (ed) Methods of enzymatic analysis: 3rd ed vol IV, Verlag Chemie, Weinheim, pp 167–177
15. Stepan J, Skrha J (1979) Measurement of amylase isoenzymes in human serum and urine using a DEAE-cellulose mini-column method. Clin Chim Acta 91: 263–271
16. Boehm-Truitt M, Harrison E, Wolf RO, Notkins AL (1978) Radioimmunoassay for human salivary amylase. Analyt Biochem 85: 476–487
17. Legaz ME, Kenny MA (1976) Electrophoretic amylase fractionation as an aid to the diagnosis of pancreatic disease. Clin Chem 22: 57–62
18. Murray RN, Fung WP, Masarei JRL, Tan EGC (1980) P3 amylase isoenzyme in the diagnosis of pancreatitis. Ann Acad Med Singapore 9: 525–528
19. LeClerc P, Forest JC (1983) Variation in amylase isoenzymes and lipase during acute pancreatitis and in other disorders causing hyperamylasemia. Clin Chem 29: 1121–1123
20. Navarro S, Aused R, Casals E, Elena M, Garcia-Puges AM, Adrian MJ, Ballesta AM, Bonet AV (1987) Value of the P3 amylase fraction as an indicator of the long-term prognosis of acute pancreatitis. Br J Surg 74: 405–407
21. Weaver DW, Bouwman DL, Walt AJ, Clink D, Sessions S, Stephany J (1982) Aged Amylase. A valuable test for detecting and tracking pancreatic pseudocysts. Arch Surg 117: 707–711
22. Warshaw AL, Lee KH (1980) Aging changes of pancreatic isoamylases and the appearance of "old" amylase in the serum of patients with pancreatic pseudocysts. Gastroeneterol 79: 1246 to 1251
23. Zakowski JJ, Bruns DE (1985) Biochemistry of human alpha amylase isoenzymes. CRC Crit Rev Clin Lab Sci 21: 283–322
24. Taussig LM, Wolf RO, Deckelbaum RJ (1974) Use of serum amylase isoenzymes in evaluation of pancreatic function. Pediatrics 54: 229–235
25. Skude G, Eriksson S (1976) Serum isoamylase in chronic pancreatitis. Scand J Gastroenterol 11: 525–527
26. Magid E, Horsing M, Rune SJ (1977) On the quantitation of isoamylases in serum and the diagnostic value of serum pancreatic type amylase in chronic pancreatitis. Scand J Gasteroenterol 12: 621–627
27. Fahrenkrug J, Magid E (1980) Concentration of immunoreactive trypsin and activity of pancreatic isoamylase in serum compared in pancreatic diseases. Clin Chem 26: 1573–1576
28. Ammann RW, Buhler L, Pei P (1982) Comparative diagnostic acuracy of four tubeless pancreatic function tests in chronic pancreatitis. Scand J Gastroenterol 17: 997–1002
29. Wolf RO, Taussig LM, Ross ME, Wood RE (1976) Quantitative evaluation of serum pancreatic isoamylases in cystic fibrosis. J Lab Clin Med 87: 164–168

30. Foo Y, Rosalki SB, Ramdial L, Mikhailidis D, Dandona P (1980) Serum isoamylase activities in diabetes mellitus. J Clin Path 33: 1102–1105
31. Skude G, Kollberg H (1976) Serum isoamylases in cystic fibrosis. Acta Paediat Scand 65: 145–149
32. Skude G (1975) Sources of the serum isoamylases and their normal range of variation with age. Scand J Gastroenterol 10: 577–584
33. Gillard BK, Cox KL, Pollack PA, Geffner ME (1984) Cystic fibrosis serum pancreatic amylase. Useful discriminator of exocrine function. Am J Dis Child 138: 577–580
34. Fridhandler L, Berk JE, Ueda M (1972) Isolation and measurement of pancreatic amylase in human serum and urine. Clin Chem 18: 1493–1497
35. Goldberg DM, Spooner RJ, Knight AH (1974) Reinterpretation of hyperamylasemia in diabetic coma. Clin Chem 20: 673–675
36. Warshaw AL, Bellini CA, Lee K-H (1976) Electrophoretic identification of an isoenzyme of amylase which increases in serum in liver disease. Gastroenterol 70: 572–576
37. Donaldson LA, Joffe SN, McIntosh W, Brodie MJ (1979) Amylase activity in bile. Gut 20: 216–218
38. Rosenblum JL, Raab BK, Alpers DH (1982) Hepatobiliary and pancreatic clearance of circulating pancreatic amylase. Am J Physiol 243: 621–627
39. Murata A, Ogawa M, Fujimoto K-I, Kitahara T, Matsuda Y, Kosaki G (1981) Hepatic implication in the metabolism of human salivary amylase isoenzyme. In: Electrophoresis '81. de Gruyter, Berlin, pp 463–470
40. Nieson TE, Alpers DH, Stahl PD, Rosenblum JL (1984) Metabolism of glycosylated human salivary amylase: In vivo plasma clearance by rat hepatic endothelial cells and in vitro receptor mediated pinocytosis by rat macrophages. J Leuk Biol 36: 307–320
41. Berk JE, Fridhandler L, Webb SF (1979) Does hyperamylasemia in the drunken alcoholic signify pancreatitis. Am J Gastroenterol 71: 557–562
42. Skude G, Wadstein J (1977) Amylase, hepatic enzymes and bilirubin in serum of chronic alcoholics. Acta Med Scand 201: 53–58
43. Rosalki SB, Nemesanszky E, Foo AY (Unpublished observations)
44. Wadstein J, Skude G (1979) Changes in amylase, hepatic enzymes and bilirubin in serum upon induction of alcohol abstinence. Acta Med Scand 205: 313–316
45. Epstein O, Chapman RWG, Lake-Bakaar G, Foo AY, Rosalki SB, Sherlock S (1982) The pancreas in primary biliary cirrhosis and primary sclerosing cholangitis. Gastroenterol 83: 1177 to 1182
46. Foo AY, Rosalki SB, Epstein O (unpublished observations)
47. Lechi A, Montesi G, Solbiati M, Rizotti P, Zanchetti M, Burlina A (1982) Serum pancreatic enzymes in acute viral hepatitis. Lab IX A23
48. Fahrenkrug J, Staun-Olsen P, Magid E (1981) Immunoreactive trypsin and pancreatic isoamylase activity in serum of patients with chronic renal failure or hepatic cirrhosis. Clin Chem 27: 1655–1657
49. Barros FP, Espinheira R, Gaeda H, de Moura MC (1986) Hyperamylasemia with an abnormal isoamylase distribution in patients with liver disease. Am J Gastroenterol 81: 261–265
50. Tsianos EB, Jalali MT, Gowenlock AH, Braganza JM (1986) Serum isoamylases in liver disease. Hepato-Gastroenterol 33: 247–249

Clinical Significance of Recent Developments in Serum Bilirubins

N. Blanckaert, J. Fevery, F. Vanstapel, and M. Muraca*

Departments of Clinical Pathology and Medicine, Catholic University of Leuven, School of Medicine, B-3000 Leuven, Belgium

Normal serum contains *unconjugated bilirubin* and small concentrations of *bilirubin glucuronides* (approximately 4% of total bilirubins). A second conjugated form of bilirubin, tentatively identified as *bilirubin-protein conjugates*, with pharmacokinetic properties that are clearly distinct from those of bilirubin glucuronides, occurs in serum of patients in whom hepatic elimination of bilirubin glucuronides is, or recently has been, impaired.

The major types of disorders of bilirubin metabolism producing hyperbilirubinemia include bilirubin overproduction, deficient hepatic clearance of the unconjugated pigment as a result of defective hepatic uptake or glucuronidation, and impaired hepatobiliary excretion of bilirubin glucuronides. Typical patterns of serum bilirubins appear to be associated with each of these broadly defined pathophysiological mechanisms and are defined by the extent to which unconjugated bilirubin and/or bilirubin glucuronides are increased in serum.

A raised ratio of unconjugated bilirubin to bilirubin glucuronides reflects impaired hepatic clearance of the unconjugated pigment. Bilirubin overproduction typically produces hyperbilirubinemia with a proportional increase of both unconjugated and glucuronidated bilirubins resulting in a ratio similar to that in normal plasma. Impaired excretory function of the hepatobiliary system primarily affects the serum level of bilirubin glucuronides and results in an increased ratio of bilirubin glucuronides to unconjugated bilirubin.

Appearance of bilirubin-protein conjugates in serum does not immediately and directly reflect hepatic function, since it appears to be the result of nonenzymic conversion of bilirubin glucuronides to protein-linked pigment in serum and occurs with some delay following a rise of serum bilirubin glucuronides. Also, disappearance of bilirubin-protein conjugates from plasma does not depend on hepatobiliary excretion, but on catabolism of the respective protein moiety which has a relatively long half-life. Therefore, conjugated hyperbilirubinemia may persist for a limited time after hepatobiliary pigment excretion has returned to normal and the serum bilirubin glucuronide concentrations have fallen.

Accurate determination of unconjugated bilirubin, and particularly of bilirubin glucuronides, requires highly specific and sensitive methodology which presently is unsuitable for routine testing in the clinical laboratory. Results of fractionation of serum bilirubins by the Hymans van den Bergh diazo method or modifications of this technique are highly procedure-dependent and may be unreliable, particularly when the level of total bilirubins is normal or only slightly raised.

* present address: Dept. of Medicine, Policlinico Universitario, Padova, Italy

Progress in Clinical Biochemistry
and Medicine, Vol. 8

1 Introduction

Physicians, veterinarians, and other bioscientists have a longstanding interest in bile pigment metabolism and analysis of bilirubin in body fluids. Detection of bilirubin in horse and human serum was first reported, respectively, in 1878 by Hammersten and in 1903 by Gilbert and co-workers. Hymans van den Berg and Snapper introduced a quantitative assay for serum bilirubins in 1913. A few years later, Hymans van den Berg and Müller discovered that the pigment in serum was heterogeneous and could be fractionated into indirect-reacting and direct-reacting bilirubins [1]. By the second decade of this century, it was appreciated that icterus could reflect either various hematologic abnormalities or hepatobiliary disorders. Ever since these classical studies, determination and fractionation of serum bilirubins has remained an indispensible diagnostic tool in clinical medicine.

The clinical importance of serum bile pigments is reflected in the large number of bilirubin measurements performed worldwide in clinical laboratories. Nonetheless, conventional methods currently in use have major limitations, in particular with respect to differentiation of the various bilirubin fractions. Hence, efforts continue to improve the analytical methodology, as evidenced by the steady stream of reports describing modifications aimed at improving the specificity of serum bilirubin differentiation. Confusion still exists with regard to the interpretation of results of serum bilirubin analyses. Many clinicians and clinical laboratorians alike are at a loss concerning the clinical significance, if any, of recent developments in the chemistry of bile pigments and new methodology introduced in some clinical laboratories for measurement of serum bilirubin. The factors mainly responsible for the present uncertainty include the lability and complex chemistry of bilirubins, the confusing terminology used to denote the various forms of bilirubin in body fluids, conflicting reports concerning certain aspects of bilirubin and pathophysiology, and the unavailability of satisfactory analytical methods in clinical laboratories.

The purpose of the present paper is to summarize and clarify the recent developments in bilirubin chemistry, metabolism, and methodology that are of importance for the diagnostic use of serum bilirubin analyses. This information should contribute to more rational utilization and interpretation of serum bilirubin analyses and reveal that the recent findings pose a number of new questions.

2 General Aspects of Bilirubin Metabolism

Unconjugated bilirubin-IXα (Fig. 1) is the principal degradation product of heme. The first step in this degradation pathway involves oxidative cleavage of the α-methene bridge and is catalyzed by microsomal heme oxygenase. In a second step, the intermediate biliverdin-IXα is transformed to bilirubin-IXα by a cytosolic bilverdin reductase. A small fraction of heme is naturally degraded by cleavage at the β-, γ-, and δ-methene bridges. This results in trace amounts of, respectively, the bilirubin IXβ, -IXγ, and -IXδ isomers. These non-α isomers can be detected in bile and/or

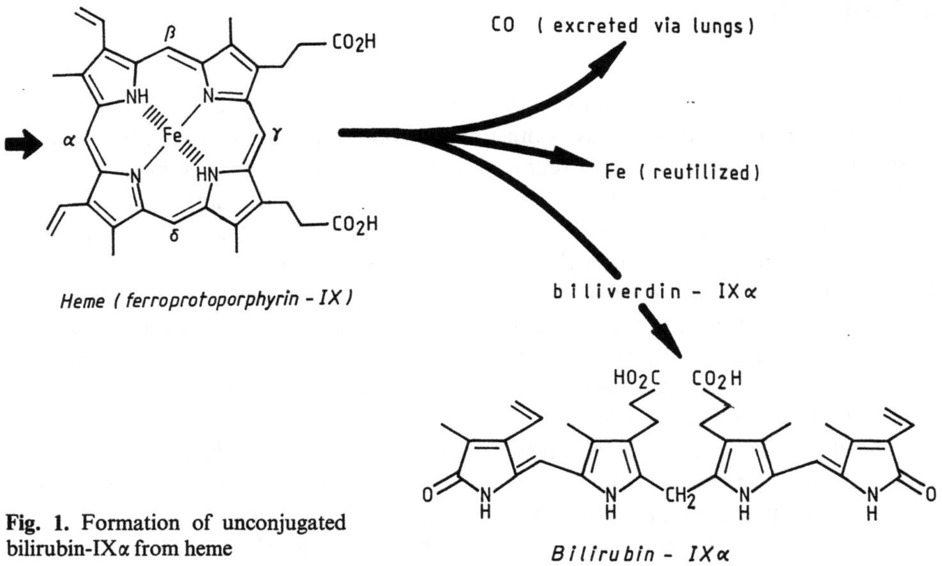

Fig. 1. Formation of unconjugated bilirubin-IXα from heme

serum when specific and sensitive methods are used [2,3]. In contrast to the major IXα isomer, bilirubin-IXβ, bilirubin-IXγ, and bilirubin-IXδ are water-soluble at physiological pH and do not require conjugation for efficient excretion in bile and urine [4].

The major source of bilirubin (approx. 65% to 80% of the total production) is the heme moiety of hemoglobin from senescent erythroid cells degraded extrahepatically, in the reticuloendothelial system. The scheme of bilirubin metabolism (Fig. 2) indicates that nearly all bilirubin produced in the body, except a part of the pigment resulting from turnover of hepatic heme compounds, enters the plasma compartment. Bilirubin in plasma is reversibly and essentially completely bound to albumin and transported to the liver, which is normally the sole organ removing bilirubin from plasma. Hepatic extraction is 2.3 + 0.4% of the plasma unconjugated bilirubin per minute, which corresponds to an uptake averaging 4.6 µg/min/kg under normal conditions. Hepatic uptake, which seems to be carrier-mediated, has a much larger capacity. About 40% of the pigment seems to reflux back unaltered from the liver into

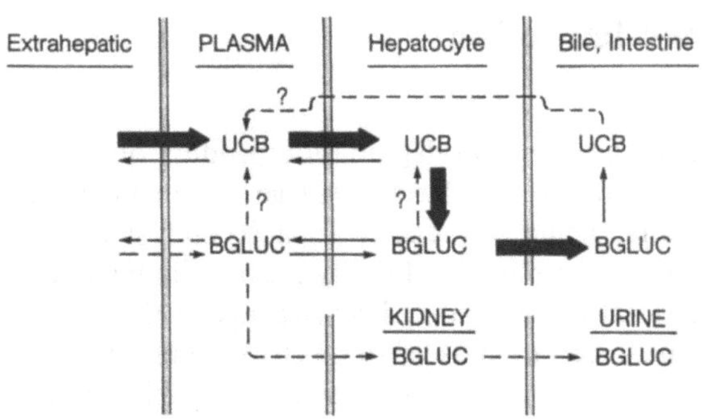

Fig. 2. Types and sources of bilirubins in normal body fluids

plasma [5]. Several excellent recent review articles are available which discuss the evidence that bilirubin is taken up in the liver by a common carrier-mediated bidirectional transport system shared, at least partially, with sulfobromophthalein (BSP), indocyanine green, and esterified bilirubins [6], but distinct from the putative carriers involved in bile acid or fatty acid uptake [7-9].

A liver plasma membrane protein that binds both BSP and bilirubin with high affinity, but not bile acids and fatty acids, has been isolated in several laboratories [7, 10, 11]. The hypothesis that this fraction may correspond to or be part of the postulated shared membrane carrier system for BSP and bilirubin is supported by the following observations. Monospecific antibodies to the purified protein, which inhibit the binding of BSP and bilirubin, are capable of impairing hepatocytic and hepatic uptake specifically of BSP and bilirubin [12]. Furthermore, BSP transport into liposomes could be reconstituted by insertion of the isolated membrane protein in phospholipid vesicles [13]. There is some kinetic evidence that hepatic uptake of BSP and bilirubin reflects not a unique carrier but several, presumably slightly different membrane proteins with overlapping substrate selectivities. For example, uptake of bilirubin and BSP are differentially affected by pretreatment with the hypolipidemic drug nafenopin in rats [14]. Moreover, there are indications that at least some of the presumed structural variants of the carriers involved in BSP and/or bilirubin uptake may be able to transport also other organic anions that are primarily transported by other, distinct carrier systems [7]. Preliminary evidence suggests that the various mono-esterified and di-esterified bilirubins share the same carrier system with unconjugated bilirubin for transport, presumably bidirectional, across the sinusoidal membrane [6]. In view of the structural differences between the various bilirubins and the above-mentioned evidence for bilirubin and BSP carrier heterogeneity, it is quite possible that the individual esterified bilirubins interact differently with the putative individual membrane carrier proteins.

It is unknown whether the macromolecular bilirubin-protein conjugates (see below) present in the circulation of patients with conjugated hyperbilirubinemia may act as "dead end" inhibitors of hepatic uptake of unconjugated bilirubin and its ester conjugates. How bilirubin is transported from the sinusoidal plasma membrane to the endoplasmic reticulum, where esterification occurs, is unknown. The classical hypothesis is that bilirubin dissociates from the putative membrane carrier protein at the internal face of the membrane and re-associates with cytosolic binding proteins, predominantly ligandin. Binding of the pigment to ligandin may facilitate intracellular transport of the endoplasmic reticulum membranes, like albumin and hemoglobin mediate facilitated diffusion of, respectively, bilirubin and oxygen in the vascular system. Efflux of bilirubin and dibromosulfophthalein from the liver into plasma shows a good correlation with hepatic ligandin concentration [15, 16]. This finding supports the view that transfer of ligand occurs between plasma membrane carriers and cytosolic ligandin.

Once in the endoplasmic reticulum, bilirubin is esterified with monosaccharides, mainly glucuronic acid, to form mono-esterified and di-esterified products. This reaction is catalyzed by a microsomal UDP-glycosyltransferase system in the endoplasmic reticulum. Conjugation of bilirubin normally proceeds well below the maximal capacity of overall transhepatic transport, which in man has been estimated to average 55.2 mg/kg/day [17].

Reports that bilirubin diglucuronide is synthesized in the canalicular plasma membrane by a distinct, so-called transglucuronidase enzyme [18] have not been confirmed despite intense investigation in several laboratories [19-21]. In fact, preliminary evidence suggests that the two isomeric natural monoglucuronides and diglucuronides are formed at a single substrate-binding site of microsomal bilirubin UDP-glucuronyltransferase [22,23].

Esterification of bilirubin with glucuronic acid, glucose, and xylose appears to be catalyzed by a single member of the microsomal UDP-glycosyltransferase system, which consists of multiple transferase isoforms with overlapping substrate selectivities toward a large number of endogenous and exogenous compounds [24,25]. The primary structure of several hepatic UDP-glucuronyltransferase isoforms has recently become available based on enzyme purification work and elegant cDNA studies [26-30]. A common feature for these enzymes is that near to the C-terminal end, they contain a segment of hydrophobic amino acid, which is typical for a transmembrane segment. The presumed transmembrane segment most likely anchors the protein in the endoplasmic reticulum membrane, with a short C-terminal portion exposed at the cytoplasmic side and a long, catalytic center-containing N-terminal portion "dangling" inside the cisternal space of the endoplasmic reticulum. If one postulates that substrate access across the endoplasmic reticulum membrane to the luminally oriented catalytic center is rate-limiting, such topological arrangement of the transferases in the endoplasmic reticulum membrane would nicely account for the finding that sealed microsomal vesicles only partially express their transferase content [31]. This so-called latency of the transferases, including the bilirubin glycosyltransferase activities, may also exist in the intact endoplasmic reticulum. Because unconjugated bilirubin diffuses rapidly into and across membranes [32], it is suggested that membrane translocation of the UDP-sugars, which are synthesized in the cytosol, limits the bilirubin esterification rate in intact, well sealed microsomes and thereby accounts for the latency phenomenon. This concept is supported by recent evidence that saturable, carrier-mediated transport of UDP-sugars occurs across the endoplasmic reticulum membrane [33,34] and that sealed rat microsomes contain an intravesicular pool of endogenous UDP-glucose [35]. However, it has been impossible so far to detect intact UDP-glucuronic acid inside the microsomes [34,35]. There presently is very little information available as to how transport of substrates to the catalytic sites of the transferases and expression of microsomal enzyme activity are regulated in the intact hepatocyte.

Both bilirubin monoglucuronides and bilirubin diglucuronide reversibly bind to ligandin, with affinity constants comparable to that for the unconjugated pigment [36]. Whether all esterified pigment, after its formation in the endoplasmic reticulum, is returned to the cytosol and becomes available for binding to ligandin and exit from the hepatocyte by carrier-mediated transport across both the sinusoidal membrane (see above) and the canalicular membrane (see below) is unresolved. There is circumstantial evidence that a small portion of the mono-esterified and di-esterified bilirubins formed in the liver normally refluxes back into plasma [37]. Bilirubin glucuronides detected in normal human serum (see below) most likely are derived from the liver because there presently is no evidence in humans for conjugation of bilirubin in extrahepatic locations.

The pathways and mechanism(s) whereby the esterified bilirubins are transported to the canalicular pole of the hepatocyte and transferred into bile are poorly understood. Current evidence indicates that passage of the esterified bilirubins across the canalicular membrane is carrier-mediated, and it is generally believed that this process is active and rate-limiting for the overall transhepatic transport of the pigment [38].

3 Structures and Origins of Serum Bile Pigments

It is now evident that a complex mixture of bilirubin pigments may occur in serum, so that the classical distinction between indirect-reacting and direct-reacting bilirubin is an oversimplification. In fact, conjugated bilirubin in cholestatic serum was found to consist of at least two different pigment species: the bilirubin sugar ester conjugates and a fraction in which the pigment is irreversibly bound to protein [39–41]. An unnecessarily complex and confusing nomenclature for bilirubins has cropped up in recent literature. The terms "delta bilirubin", "bil-alb", "biliprotein", and "protein-bound bilirubin" have been used to denote the serum bilirubin fraction tentatively identified as bilirubins covalently bonded to serum protein, while "conjugated bilirubin" is frequently used to refer to the bilirubins esterified with one or two sugar moieties. This terminology is unsatisfactory for a number of reasons. For example, "delta bilirubin" is totally nondiscriptive. Conjugation generally refers to covalent linkage of an endogenous compound with another, either endogenous or exogenous substance. Therefore, both the bilirubins esterified with one or two sugars and the bilirubins covalently linked to plasma protein are properly designated by the term "bilirubin conjugates". "Protein-bound" bilirubin cannot specifically refer to bilirubin-protein conjugates because the unconjugated and esterified pigments engage in tight, reversible binding to albumin and ligandin.

We propose the term "bilirubin-protein conjugate" to specifically denote the pigment that is covalently bonded to serum protein [42]. Two types of natural bilirubin-protein conjugates exist. An esterified bilirubin-protein conjugate corresponds with pigment that is covalently bonded to protein (probably at a propionic acid side-chain) and that is also esterified, at the other propionic acid side-chain, with a sugar moiety. An unesterified bilirubin-protein conjugate consists of bilirubin with one free propionic acid side-chain and another propionic acid side-chain linked to protein. We use "esterified bilirubins" to denote bilirubins in which one of the two propionic acid side-chains are esterified and covalent linkage of the pigment to protein is absent. The term "conjugated bilirubins" comprises mono- and di-esterified bilirubins, as well as the two types of bilirubin-protein conjugates described above.

Natural bile pigments include the unconjugated bilirubin-IXα, bilirubin-IXβ, bilirubin-IXγ and bilirubin-IXδ isomers and the mono- and di-esterified bilirubins. Glucuronosyl, glucosyl and xylosyl moieties are the conjugating groups found in natural esterified bilirubins. During cholestasis in vivo and probably also in vitro, esterified bilirubins can nonenzymically react with plasma proteins, mainly albumin, and thereby be transformed into bilirubin-protein conjugates. In addition, both in vitro and in vivo stasis may lead to transformation of the enzymically synthesized

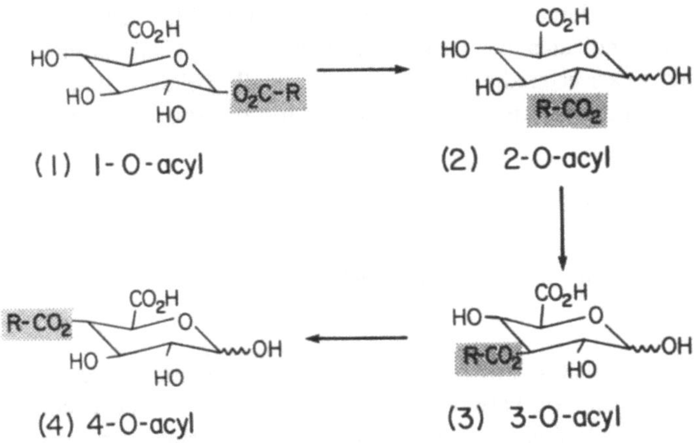

Fig. 3. Structures of four positional isomers of bilirubin glucuronide formed by migration of the bilirubin acyl group (R) from the C-1 position of the glucuronosyl moiety to C-2, C-3, and C-4

bilirubin 1-O-acyl ester conjugates to esterified bilirubins linked at the C-2, C-3 or C-4 atom of the glycosyl group (Fig. 3), as a result of nonenzymic migration of the bilirubin acyl group from the original C-1 atom of the sugar moiety to the C-2, C-3, and C-4 positions [43,44]. Phototherapy results in formation of a number of conformational isomers of unconjugated bilirubin-IXα and so-called lumirubins. Measurement of these photobilirubins and other photooxygenation products possibly formed during phototherapy requires special techniques and is not of obvious clinical importance. Extensive discussion of phototherapy aspects can be found in several recent papers and reviews by McDonagh and Lightner [45-47]. Bilirubin-IXα, the major physiologically important pigment exhibits two propionic acid side chains (Fig. 1). These functional groups are the sites of conjugation in the natural mono-esterified and di-esterified bilirubins. The free propionic acid side-chains probably are largely ionized in vivo. The quoted pK_a values vary from 4.4 till 6.9 although most authors favour 6.2–6.5 [48]. The lactam groups of the two lateral rings are also slightly basic and can be protonated in an acidic medium. X-ray diffraction studies of the crystalline structure have demonstrated that the polar groups are involved in multiple intramolecular hydrogen bonds [49], which confer to the molecule a ridge tile structure. The intramolecular H bonds limit the possibility of forming H bonds with water, thus explaining the low solubility (7 n*M* at pH 7.4) of the pigment in aqueous fluids [50]. Solubilization of unconjugated bilirubin-IXα in body fluid bilirubin is effected by tight, reversible binding to carrier proteins, predominantly albumin in plasma and ligandins in cytosol. Conjugation with one or two sugar residues prevents formation of internal hydrogen bonds due to sterical hindrance. The bilirubin-IXβ, bilirubin-IXγ, and bilirubin-IXδ isomers cannot form intramolecular hydrogen bridges, which presumably explains their hydrophilicity. Furthermore, non-α isomers of unconjugated bilirubin can be excreted in bile without conjugation [4]. The involuted, hydrogen-bonded structure of unconjugated bilirubin can also explain the "indirect" diazo-reaction of the unconjugated pigment, since the attack of the diazo-reagent at the central methylene bridge is very probably hindered by the intramolecular hydrogen bonds. It is speculated that the so-called accelerator substances used in the diazo reaction exert their effect by disrupting the intra-molecularly hydrogen-bonded forms of bilirubin-IXα, so that the pigment molecule becomes flexible and can effectively be attacked at its central methylene bridge by the diazo-reagent.

Bilirubin mono-ester and di-ester conjugates are synthesized in the liver by enzymic esterification of one or both of the propionic acid groups of the tetrapyrrole with a sugar moiety. Linkage of the conjugating group in these esters is in the β-D configuration at the C-1 position of the sugar (Fig. 3). The ester bond is labile, and it can undergo a series of modifications.

Recent evidence indicates that three types of modifications occur to esterified bilirubins that accumulate in serum of patients with impaired hepatobiliary excretion. First, the ester linkage in the bilirubin sugar conjugates is alkali-labile which probably leads to easy nonenzymic hydrolysis even at the slightly alkaline physiologic pH in the circulation [51]. Such nonenzymic saponification possibly together with β-glucuronidase-mediated hydrolysis may at least partially explain why there is an enrichment of the mono-esterified and unconjugated pigment fractions in serum in comparison with bile in patients with conjugated hyperbilirubinemia [2,42]. Second, the bilirubin acyl group in the esterified pigment readily shifts, by a nonenzymic intramolecular mechanism, from the original carbon-1 position of the conjugating sugar moiety consecutively to the carbon-2, -3, and -4 positions (Fig. 3). Thus, each of the already large number of different C-1-linked bilirubin esters can be transformed to three additional positional (C-2, C-3, or C-4) isomeric forms [43,44,51,52]. Such "acyl-shifting" occurs in vivo, in the circulation, and results in an enormous number (theoretically over 200!) of structurally and isomerically different esterified bilirubins. It is evident that this complexity may render interpretation and calibration of chromatographic separations obtained by high-resolution methods such as high-pressure liquid chromatography next to impossible if the native underivatized pigments are chromatographed. Third, it recently became recognized that accumulation of esterified bilirubins, but not of unconjugated bilirubin in serum results in transformation to bilirubin-protein conjugates, in which the pigment moiety presumably is covalently linked to serum proteins, predominantly albumin [40–42,51–56]. The reaction mechanism by which these bilirubin-protein conjugates are formed and the nature of the linkage between protein and pigment remain unresolved. Considering the properties of the ester linkage of the bilirubin sugar conjugates, it appears likely that the formation of bilirubin-protein conjugates involves attack by (a) nucleophilic group(s) in albumin, i.e., the free sulfhydryl group, resulting in displacement of the sugar moiety and formation of a bilirubin-protein adduct by transacylation. Such a presumed transacylation mechanism is completely analogous to the previously described sugar displacement reactions for bilirubin-sugar conjugates involving attack by other non-protein nucleophiles such as water, ammonia [57,58], methanol [59,60] or (intramolecularly) by a neighboring hydroxyl group in the sugar moiety [43]. Both the glycosidic (i.e., C-1-linked) and the nonglycosidic (C-2-, C-3-, and C-4-linked) bilirubin sugar conjugates appear to react with albumin and form bilirubin-albumin conjugates (Blanckaert N, unpublished observations).

4 Determinants of Levels of Bilirubins in Serum

4.1 *Unconjugated Bilirubin*

The postulated sources of the serum bilirubins in normal serum are depicted in Fig. 2. According to this scheme, the major determinants of the unconjugated bilirubin level are (a) plasma bilirubin turnover (PT_{UCB}) and (b) the plasma bilirubin clearance (G_{UCB}), which occurs almost solely in the liver. Because almost all bilirubin is produced extrahepatically and enters the plasma compartment, PT_{UCB} directly reflects the rate of bilirubin production. Plasma influx of unconjugated bilirubin may occur by reabsorption of unconjugated pigments from the gut and hydrolysis of bilirubin esters in plasma, but these routes are probably negligible under normal conditions. C_{UCB} is determined by pigment binding to protein in plasma and cytosol, and by the properties of both the postulated microsomal membrane carrier system and the microsomal enzyme system for bilirubin esterification. Unlike for bile acids, the extraction ratio is low, which implies that liver parenchymal function and not hepatic blood flow is the primary determinant of hepatic uptake of unconjugated bilirubin.

Under steady-state conditions, the plasma unconjugated bilirubin concentration (P_{UCB}) is directly related to bilirubin production rate (BRP) and inversely to hepatic clearance (C_{UCB}), according to the equation:

$$P_{UCB} = BRP/C_{UCB}$$ [61]

For a given rate of bilirubin production, plasma unconjugated bilirubin concentration and hepatic bilirubin clearance are related to each other in a hyperbolic fashion. The equation explains why the distribution of plasma bilirubin in the population is not symmetrical about the mean, but skewed towards the higher values. The upper limit of normal serum bilirubin values in adults is generally set at 1 mg/dl or 17 μmol/l, irrespective of gender. However, to define normal levels one should also take into account that a sex difference in serum bilirubin concentration is well documented. Males having significantly higher total bilirubin values than females [62–65]. It is surprising that such a well-established observation has never been applied to clinical practice. It is not clear whether the lower total bilirubin values for serum of females are the result of a lower bilirubin production rate or of a higher hepatic clearance compared to males. The finding of a higher excretion of fecal urobilinogen in males [66] suggests a sex difference for bilirubin production rate, even if no difference was observed between males and females in plasma turnover studies with radiolabeled bilirubin [67]. Experimental work in rats has demonstrated a more efficient hepatic conjugation and lower serum bilirubin values in female animals compared to males [3,68,69].

A number of physiological, pathological and pharmacological conditions associated with increased pigment production or reduced hepatic clearance (Table 1) can lead to predominantly unconjugated hyperbilirubinemia.

Table 1. Causes of hyperbilirubinemia

Predominantly unconjugated hyperbilirubinemia

A. *Overproduction*
  Hemolysis
  Ineffective erythropoiesis
  Fasting?
B. *Decreased hepatic uptake*
  Drugs (e.g., flavaspidic acid, rifamycin, contrast media)
  Fasting
  Sepsis
  Gilbert's syndrome?
C. *Decreased bilirubin conjugation*
  Gilbert's syndrome
  Crigler-Najjar disease
  Neonatal jaundice
  Acquired conjugation defect:
  drug inhibition (e.g. pregnanendiol, chloramphenicol)
  hepatocellular disease (chronic persistent hepatitis, Wilson)
  sepsis

Predominantly conjugated hyperbilirubinaemia

A. *Intrahepatic biliary obstruction*
  Hepatocellular disease (viral or drug-induced)
  Drug-induced cholestasis (e.g., steroids)
  Sepsis
  Recurrent (benign) intrahepatic cholestasis
  Cholestatic jaundice of pregnancy
  Dubin-Johnson syndrome
  Rotor syndrome
B. *Extrahepatic biliary obstruction* (e.g., stones, tumors, strictures)

4.2 Bilirubin Ester Conjugates

Small amounts of bilirubin ester conjugates are found in serum of normal healthy adults [3,37,62]. Individual concentrations of mono-esterified and di-esterified bilirubins averaged 0.10 and 0.11 µmol/L, respectively, corresponding to 3.6% of total bilirubin in a series of 43 healthy men and women [3]. The esterified pigment in normal serum has been structurally identified as 1-O-acyl glucuronides, and are likely derived from hepatic reflux into plasma, since in humans the liver seems the only organ which can esterify bilirubin to a significant extent and since the compounds are not reabsorbed from the intestine [70]. Studies based on infusion of unconjugated bilirubin in bile-fistula rats have shown that the concentration of esterified pigments in plasma increases linearly with the infusion rate, suggesting that an equilibrium exists between the amount of conjugates formed in the liver and their plasma concentration [71]. The observed increase in esterified pigment concentration was linear even at higher infusion rates resulting in near-saturation of trans-hepatic transport. Based on these observations, it seems unlikely that the presence in plasma of esterified bilirubin reflects a relative deficiency of the bile secretory step in normal liver, as previously has been reported [72,73]. The most recent data thus suggest that, in the absence of hepatobiliary disease, the serum concentration of esterified bilirubins should directly reflect PT_{UCB} and the hepatic bilirubin esterification rate, and should be inversely related to the clearance of the esterified bilirubins from the intrahepatic compartment.

Another major determinant of the serum level of esterified pigment must be the plasma clearance of the individual esterified bilirubins. Both the liver and the kidney are potential excretory routes for esterified pigment albeit that renal elimination is negligible under normal circumstances, reflecting the high efficiency with which esterified bilirubins are extracted from plasma and excreted by the liver [21,36,74].

In serum of patients with biliary obstruction, glucose and xylose ester conjugates have been identified in addition to the more important glucuronides and their positional C-2, C-3, C-4 isomers. Enhanced relative amounts of esterified bilirubin in plasma are typically observed in hepatocellular or cholestatic liver disease, as listed in Table 1.

4.3 Bilirubin-protein Conjugates

A bilirubin fraction irreversibly and presumably covalently bound to albumin and possible to a less extent to other proteins has been characterized in serum of patients with liver diseases [39–42,51,54]. It has been demonstrated that esterified bilirubin is the precursor involved in the nonenzymic formation of the irreversible bond with the protein [40,41,51,55]. This new species of bilirubin conjugate retains its "direct-reacting" properties in the diazo-reaction [40,41]. The bilirubin-protein conjugate appears in serum in all conditions causing an accumulation of esterified bilirubin, such as biliary obstruction, hepatitis, cirrhosis, infiltrative liver disease, Dubin-Johnson syndrome and in jaundice associated with sepsis [42,54]. The process seems nonspecific and time-dependent, to the extent that the proportion of bilirubin-protein conjugate in serum tends to increase in the course of the disease [54,75]. This bilirubin species does not undergo glomerular filtration and exhibits a half-life similar to albumin. The pigment's long half-life readily explains the persistence of high serum bilirubin levels in the absence of bilirubinuria during the resolution phase of liver disease. In fact, after a desobstructive procedure, almost all of the direct-reacting bilirubin in serum consists of bilirubin-protein conjugates [54,75].

5 Methodological Aspects

5.1 Methodology Applicable to Routine Clinical Laboratories

Most current methods for the measurement of bilirubins in biological fluids are still based on the diazo-coupling of the pigments as originally described in 1913 by Hymans van den Bergh and Snapper [76]. The reaction is based on a nucleophilic attack of a first diazo-reagent molecule at C-9 or C-11 of the tetrapyrrole, with formation of a first azopyrromethene reaction product and an intermediate product which is unstable and reacts with another diazo-reagent molecule, yielding a second

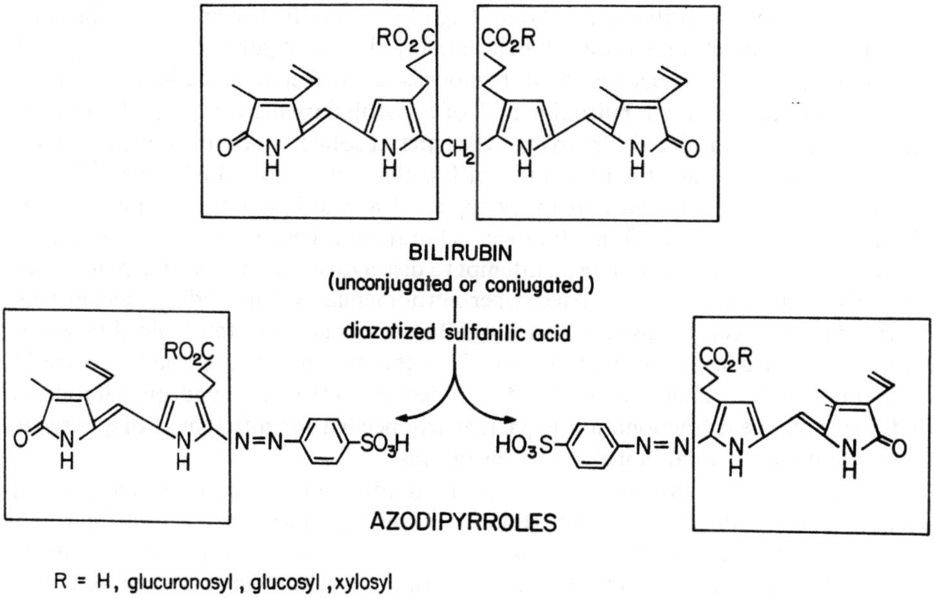

Fig. 4. Diazo-cleavage reaction of bilirubin resulting in formation of two dipyrrolic fragments

azopyrromethene derivative. Thus, in this two-stage reaction, two azopyrromethene molecules are formed from each bilirubin molecule (Fig. 4). The amount of azoderivatives formed in the reaction is determined spectrophotometrically. Hymans van den Bergh & Muller [1] observed that bilirubin in bile reacts "directly" i.e., without requiring the presence of an accelerator, such as alcohol, and concluded that at least two species of bilirubin were present in biological fluids. A direct reaction was obtained with bile and with the bilirubin in urine and plasma of patients with obstructive jaundice, while the bilirubin in plasma of patients with hemolytic disease required the presence of alcohol to give a positive diazo-reaction ("indirect-reacting" bilirubin). The "indirect-reacting" fraction was found to correspond to unconjugated bilirubin, and the "direct-reacting" pigment was later identified as bilirubin glucuronide [77–79]. More recently, it has been recognized that bilirubin-protein conjugates are direct-reacting and may constitute the major portion of the direct-reacting pigment in cholestatic serum. Several modifications of the original diazo method have been proposed (see e.g., the evaluation of analytical variables for different diazo methods in [80]). The method of Jendrassik and Grof [81] has been recommended by the National Committee for Clinical Laboratory Standards. Although the distinction between "direct" and "indirect" reacting bilirubin was found to be very useful in clinical practice, it is now clear that the two pigment fractions do not reflect the actual concentrations of unconjugated and conjugated bilirubin, but that they are influenced by the reaction conditions (such as pH or exposure to light) and by the composition of the sample [82]. In general, there is always a variable fraction of the unconjugated pigment in plasma that behaves as "direct" reacting, so that it is impossible to determine the exact concentration of esterified bilirubin in the presence of a large excess of unconjugated pigment [58]. On the other hand, it has been recognized that the diazo-method quantifies only about 70% of the esterified pigment [83]. The fractionation of serum bilirubin is traditionally employed to distinguish the "unconjugated hyperbilirubinemia" associated with hemolysis or defective hepatic conjugation from the "conjugated hyperbilirubinemia" of hepatobiliary disease. Hyperbilirubinemia is generally considered "indirect" when the absolute concentration of "direct-reacting" bilirubin is less than 5 μmol/L or less than 20% of total bilirubin [84-86].

The limitations of the diazo methods appear clearly at low concentrations of total bilirubin. In one study, four different laboratories performed serum bilirubin fractionation in a series of sera in an attempt to distinguish patients with hepatobiliary disease from those with unconjugated hyperbilirubinemia. A marked disagreement was observed for sera containing less than 5 mg/dl (or 85 μmol/L) of total bilirubin. Since serum bilirubin does not exceed 70 μmol/L in the presence of chronic hemolysis [5] and similar concentrations are observed in subjects with Gilbert's syndrome, it follows that serum pigment fractionation is particularly needed for differential diagnosis at these relatively low total bilirubin concentrations.

A significant improvement in terms of standardization and accuracy has recently been obtained with the Ektachem (R) dry chemistry slides. In the total bilirubin slide [87], diazo-coupling proceeds on a buffered porous layer and the azoderivatives become bound to a cationic polymeric mordant, resulting in a spectral change. Measurements are then made at two different wavelengths, which permits correction for spectral interferences. The test has the advantage of well controlled, reproducible reaction conditions which are often not obtained with the classical diazo methods, even when automated systems are used [88-89]. A novel approach for the differential determination of unconjugated and esterified bilirubin is proposed with the "BuBc" Ektachem slide [91]. The macromolecular bilirubin-protein conjugate is prevented by size-exclusion from migrating through the first porous spreading layer, which also traps interfering substances such as hemoglobin, lipids and lipochromes. Unconjugated bilirubin and esterified bilirubins reach the underlying registration layer where they bind to a mordant. Separate determination of the two pigment fractions is based on the different absorption spectra of the unconjugated and esterified pigment fractions when bound to the mordant. The system can be calibrated with taurine-conjugated bilirubin as a surrogate for esterified bilirubin. It has been claimed that esterified bilirubin as measured with this assay would be a better indicator of deranged hepatobiliary function than direct bilirubin [92]. However, more studies are needed to determine the sensitivity and reliability of the method to measure the concentration of the esterified pigment at low levels of total bilirubin.

5.2 Analysis of Tetrapyrroles

The limitations of conventional diazo methods stimulated the development of several procedures for the analysis of intact tetrapyrroles. Solvent partition techniques were developed in an attempt to separate the unconjugated from the diazotized conjugated pigments based on their different polarity [93], but these procedures were not reliable nor more selective than diazo methods [94]. Early chromatographic techniques were unpractical and not quantitative [94]. Recently, extraction of intact tetrapyrroles with a chloroform-ethanol mixture followed by thin-layer chromatography allowed semiquantitative determination of the different bilirubins in bile, urine, serum, stools, and liver homogenate [2]. However, the most significant progress was obtained by the utilization of high-performance liquid chromatography (HPLC). The high-performance liquid chromatograph is a closed system, which is ideal to prevent light and oxygen-induced degradation of bile pigments. Additional advantages are the short duration of chromatography, the minimal amount of sample required, the high sensitivity due to low instrumental band spread and improved detector technology, and the highly reproducible and accurate quantitation of individual bile pigments, especially when an internal standard is included in the procedure. However, most of the proposed HPLC separation systems of intact tetrapyrroles have not been sufficiently validated as analytical tools, and while some of them can be used for the analysis of bile [45,95,96], they are not adequate for serum samples.

5.3 The Alkaline Methanolysis Procedure

The alkaline methanolysis of bile pigments [60,97] allows accurate and sensitive determination of individual concentrations of unconjugated bilirubin and of esterified bilirubin in serum and bile. Bilirubin mono- and disugar conjugates are converted to the corresponding mono- and dimethyl esters in alkaline methanol. Unconjugated bilirubin is not affected by the transesterification reaction, and is extracted into chloroform together with the methyl ester derivatives. The pigments can then be separated and individually quantified by thin-layer chromatography or by HPLC. In the latter case, a known amount of xanthobilirubic acid methyl ester is mixed with the sample at the beginning of the procedure and used as internal standard for quantitation of individual pigments [3,97,98]. Since conversion of the sugar conjugates to methyl esters is virtually quantitative, this method can be calibrated with available bilirubin mono- and dimethyl esters. The originally developed HPLC separation system allowed accurate quantitation of unconjugated bilirubin and bilirubin mono- and diester conjugates in sera of liver patients with hyperbilirubinemia, but it was not sensitive enough to detect the esterified pigments in serum of healthy adults and of subjects with Gilbert's syndrome [97,99]. This was due mainly to interference with carotenoids. Later on, a reverse-phase separation was developed, employing a mobile phase in which carotenoids are insoluble [3]. The latter procedure allowed for the first time individual quantitation of the bilirubin ester conjugates in serum of healthy adults, newborns, and patients with Gilbert's syndrome, Crigler-Najjar disease, or hemolysis [3,37,100]. The assay is linear down to concentrations as low as 0.03 μM and 0.02 μM for, respectively, mono-esterified and di-esterified bilirubins.

5.4 Analysis of Bilirubin-protein Conjugates

In conjugated hyperbilirubinemia, the sum of individual concentrations of unconjugated bilirubin and of the ester conjugates measured by alkaline methanolysis-HPLC is lower than the concentration of total bilirubin determined by diazo methods [53,75,97,101]. The protein precipitate discarded during the procedure is yellow when cholestatic serum is assayed, and it contains large amounts of diazo-positive material, constituted by bilirubin-protein conjugates. This bilirubin fraction can be separated by affinity chromatography on albumin immobilized on Sepharose 4B [41] or by HPLC [39]. However, both procedures are complex and have not been extensively investigated as analytical tools. Measurement of serum bilirubin fractions by commercially available dry chemistry slides with the Kodak Ektachem analyzer permits indirect estimation of bilirubin-protein conjugates [102]. A simple method to directly assess this pigment fraction has recently been validated [42]. Reversibly-bound bilirubins are selectively removed from serum proteins by organic solvent extraction and bilirubin-protein conjugates are directly assayed by diazo-coupling on the protein pellet. The assay requires only common laboratory equipment, and it exhibits good characteristics of accuracy, precision, linearity and sensitivity. Quantitation of bilirubin-protein conjugates in serum has also been obtained with the use of C-8 reverse-phase cartridges, followed by diazo-coupling [103] or by ion-exchange chromatography of the azodipyrrole derivatives [104].

6 Patterns of Serum Bilirubins in Health and Disordered Bilirubin Metabolism

6.1 *Healthy Adults*

A consistent finding in several surveys is that the serum concentration of total bilirubins (TB; the sum of unconjugated and all conjugated bilirubins) is higher in adult males than in adult females. This sex difference arises only after puberty, when mean TB levels in females appear to decrease [65]. The same phenomenon is found in adult rats [3], and it has been postulated that the sex difference reflects more efficient hepatic bilirubin conjugation in females than in males. Despite the fact that the sex difference is well documented, virtually all clinical laboratories continue to list the same "normal" values for both sexes. It is clinically important to realize that a marginally elevated TB value is more likely to reflect disordered bilirubin metabolism in adult females than in males. Until recently, there was no assay with the sensitivity and specificity needed for accurate differential determination of unconjugated bilirubin and particularly its conjugates in serum. Methodology with such a capability has only recently been developed [3]. The preliminary results, obtained in a relatively small number of individuals, indicate that unconjugated bilirubin accounts for approximately 96% of all bilirubin pigment in serum of healthy adults. The remaining pigment consists of bilirubin mono-1-O-acyl glucuronide and bilirubin di-1-O-acyl-glucuronide in about equal concentrations, and trace amounts of a pigment tentatively identified as bilirubin-IXβ. Bilirubin-protein conjugates were undetectable (<1 μmole/l or 0.06 mg/dl) in normal human serum [42]. Trace amounts of bilirubins, undetectable by methods used in the clinical laboratories, are present in normal urine.

6.2 *Differential Diagnosis of Hyperbilirubinemia*

Icterus, defined as clinically obvious yellow discoloration of the skin and mucosae is usually recognized when serum bilirubin exceeds 2–3 mg/dl. The first diagnostic step when approaching an icteric patient is to identify the hyperbilirubinemia as "indirect-" or "direct-reacting", in order to distinguish patients with hemolytic disorders or deficient hepatic conjugation from patients with hepatobiliary disease. Serum bile acid determination may be useful for such differential diagnosis and is a sensitive liver test in patients with cirrhosis [105], but may give normal results in liver disease without cirrhosis [75]. Since, as previously observed, "direct-reacting" bilirubin does not reliably distinguish between conjugated and unconjugated hyperbilirubinemia when total serum bilirubin is only moderately elevated, fractionation of serum bilirubin even has been abandoned by some clinical laboratories. However, preliminary results obtained with the alkaline methanolysis-HPLC assay suggest that accurate, separate measurements of unconjugated bilirubin and of bilirubin ester conjugates may be more reliable for differential diagnosis in patients with hyperbilirubinemia. Increased levels of conjugated pigments can be demonstrated in patients with liver disease even at normal levels of total bilirubin [75,99] and preliminary observations suggest that the ratio of esterified over total bilirubin is specifically increased in hepatobiliary disorders [75]. Moreover, it was demonstrated that unlike in Gilbert's syndrome or Crigler-Najjar disease, patients with hyperbilirubinemia due to bilirubin overproduction display a parallel increase of both unconjugated and conjugated bilirubins (Fig. 5) [37]. In summary, the three fundamental alterations of bilirubin metabolism are reflected by three different patterns of serum bilirubins, which can be differentiated by alkaline methanolysis-HPLC [37].

1) The purely unconjugated hyperbilirubinemia associated with deficient hepatic conjugation of unconjugated bilirubin with normal or decreased absolute concentrations of the ester conjugates, resulting in a decreased percentage of esterified versus total bilirubin.

2) A parallel increase of both unconjugated and esterified bilirubins, with a normal ratio of esterified to total pigment, is characteristic of increased pigment production.

3) A normal or a moderately increased unconjugated bilirubin with a markedly increased concentration of esterified bilirubin, resulting in an increased percentage of esterified to total pigment, is found in hepatobiliary disease.

Further differentiation of esterified bilirubins into mono- and diester conjugates does not provide additional clinically useful information. The ratio of mono- to diester conjugates ratio is increased both in hemolytic jaundice and in Gilbert's syndrome [37]. The ratio seems also higher in patients with hepatocellular disease than in those with extrahepatic obstruction, probably reflecting prolonged stasis in the hepatocyte [106], but considerable overlap exists between the two groups [75,101].

Just as the other liver tests, serum bilirubin cannot discriminate between intrahepatic and extrahepatic biliary obstruction. In patients with cholestasis, a discrepancy may be observed between serum bilirubin and the other liver tests. Total serum bilirubin is usually normal in the case of partial or incomplete biliary obstruction, while serum bile acids and alkaline phosphatase are markedly elevated. Such a discrepancy reflects the different mechanisms which determine the concentrations of these compounds in serum of cholestatic patients. In case of a localized biliary obstruction, as can occur with intrahepatic gallstones or tumors or with sclerosing cholangitis or primary biliary cirrhosis (early phase), the large reserve capacity of the non-obstructed parenchyma to excrete bilirubin may explain why the total bilirubin concentration remains within normal limits. On the contrary, serum concentrations of alkaline phosphatase and bile acids are elevated in such conditions, apparently as a result of enzymatic induction and altered permeability of the biliary tree [107].

The "physiological" jaundice of the newborn deserves separate discussion. Neonatal hyperbilirubinemia is the result of both increased pigment production and decreased hepatic clearance during the first days of life. Large amounts of unconjugated bilirubin accumulate in blood, reaching a maximum around the fifth day after birth and declining thereafter [108]. The appearance of esterified bilirubin in neonatal serum has long been questioned, and was considered either as an index of maturation of hepatic glucurono-conjugation [109] or as a sign of inefficient biliary excretion [110-112]. However, these conclusions were drawn from results obtained with methods unreliable to quantify the conjugates. The existence of a state of "physiological cholestasis" in neonatal life has been put forward as high concentrations of serum bile acids are present in newborns [113-115]. However, this finding does not necessarily indicate an immaturity of biliary excretion. In fact, a significant accumulation of esterified bilirubin could not be demonstrated with alkaline-methanolysis-HPLC analysis of neonatal serum [100], despite the fact that the newborn liver is eliminating a load of pigment which is much higher than occurring in adults. Bilirubin ester conjugates are present in neonatal serum, but, even if their absolute concentration is higher than in adults, they constitute only about 1% of total pigment. This pattern is the result both of the increased pigment production and of the decreased hepatic conjugation characteristic of neonatal life in humans. The situation is quite different in newborn rat serum, where the concentration of esterified bilirubin rises steeply at birth to account for approximately 40% of the total pigment [116]. This observation is in agreement with a neonatal immaturity of biliary secretion present in this species, as shown also by electron microscopy [117].

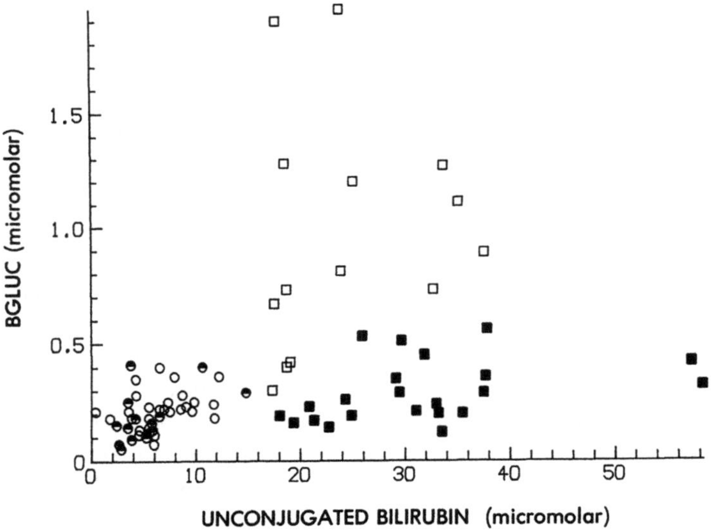

Fig. 5. Relationship between the serum concentrations of unconjugated bilirubin and esterified bilirubins in healthy women (◒) and men (○), patients with Gilbert's syndrome (■), and patients with hemolytic disease (□). BGLUC, sum of monoesterified and diesterified bilirubins

6.3 Serum Bilirubin as a Screening Test

To be used for screening, a test should be simple, highly sensitive, and reasonably specific. Total serum bilirubin measured by a diazo method is easy to perform and can be automated. Its sensitivity, however, for detection of mild parenchymal or biliary liver diseases such as chronic hepatitis, inactive cirrhosis, early sclerosing or destructive non-suppurative cholangitis (early state of primary biliary cirrhosis) is low. Moreover, the test lacks specificity, as it is elevated also in hematological conditions and in subjects with congenitally deficient hepatic conjugation. Detection of enhanced concentrations of esterified bilirubin in serum is probably a more specific and sensitive indicator of liver disease, since this pigment derives only from hepatic reflux and since biliary secretion of bilirubin is sensitive to a variety of insults. Elevation of direct-reacting bilirubin in the presence of normal total bilirubin has been reported in serum of patients with acute and chronic hepatocellular disease, congestive heart failure and hepatic neoplasma. As mentioned above, the so-called "direct-reacting" fraction is not specific for the conjugated pigment at low total bilirubin levels, but these results seem to be confirmed by the finding of increased concentrations of bilirubin ester conjugates as measured by the alkaline methanolysis-HPLC procedure in 89 out of 91 patients with various forms of liver disease [97], in 15 out of 28 serum samples from patients with liver metastases and normal total bilirubin levels, and in 14 out of 22 patients with liver cirrhosis and normal or slightly increased total bilirubin [99]. In another group of 148 liver patients, 97% exhibited enhanced serum concentrations of conjugated bilirubins, determined by the same method [75]. This test was more sensitive than any of four other liver tests namely serum bile acids, γ-glutamyl transpeptidase, alkaline transferase, and alkaline phosphatase, and was as sensitive as a combination of the four tests. In the same series, esterified bilirubin was elevated in 15 out of 18 patients with normal total bilirubin levels. The above-mentioned results were all obtained with the normal-phase chromatographic system, which permits detection of ester conjugates in serum only when they are increased beyond an ill-defined concentration. Application of the more recent reverse-phase method [3], which can quantify esterified bilirubin in serum of healthy individuals and thus allows to establish the normal range of values, should improve both the sensitivity and the specificity of the assay in liver patients. Unfortunately, such highly sensitive and specific methods for measurement of bilirubin conjugates cannot yet be automated and, therefore, cannot be used routinely at present.

7 References

1. Hymans van den Bergh AAH, Müller P (1916) Über eine direkte und eine indirekte Diazoreaktion auf Bilirubin. Biochem Z 77: 90
2. Fevery J, Blanckaert NN, Leroy P, Michiels R, Heirwegh KPM (1983) Analysis of bilirubins in biological fluids by extraction and thin-layer chromatography of the intact tetrapyrroles: application to bile of patients with Gilbert's syndrome, hemolysis or cholelithiasis. Hepatology 3: 177

3. Muraca M, Blanckaert N (1983) Liquid chromatographic assay and identification of mono- and diester conjugates of bilirubin in normal serum. Clin Chem 29: 1767
4. Blanckaert N, Heirwegh KPM, Zaman Z (1977) Comparison of the biliary excretion of the four isomers of bilirubin-IX in Wistar and homozygous Gunn rats. Biochem J 164: 229
5. Berk PD, Martin JF, Blaschke TF, Scharschmidt BF, Plotz PH (1975) Unconjugated hyperbilirubinemia-physiologic evaluation and experimental approaches to therapy. Ann Int Med 82: 552
6. Shupeck M, Wolkoff AW, Scharschmidt BF, Waggoner JG, and Berk PD (1987) Studies of the kinetics of purified conjugated bilirubin-^{3}H in the rat. Am J Gastroenterol 70: 259
7. Berk PD, Potter BJ, Stremmel W (1987) Role of plasma membrane ligand-binding proteins in the hepatocellular uptake of albumin-bound organic anions. Hepatology 7: 165
8. Wolkoff AW (1987) The role of an albumin receptor in hepatic organic anion uptake: The controversy continues. Hepatology 7: 777
9. Sorrentino D, Stremmel W, Berk BD (1987) The hepatocellular uptake of bilirubin: current concepts and controversies. Molec Aspects Med 9: 405
10. Wolkoff AW, Sosiak A, Greenblatt HC et al (1985) Immunological studies of an organic anion binding protein (OABP) isolated from rat liver cell plasma membranes. J Clin Invest 76: 454
11. Tiribelli C, Lunazzi GC, Sottocase GL (1986) Mechanisms of hepatic uptake of organic anions. Clin Sci 7: 1
12. Stremmel W, Berk PD (1986) Hepatocellular uptake of sulfobromophthalein (BSP) and bilirubin is selectively inhibited by an antibody to the liver plasma membrane BSP/bilirubin binding proteins. J Clin Invest 78: 822
13. Sottocasa GL, Baldini G, Sandri G, et al (1982) Reconstitution in vitro of sulfobromophthalein transport by bilitranslocase. Biochim Biophys Scta 685: 123
14. Gartner V, Stockert RJ, Levine WG et al (1982) Effect of nafenopin on the uptake of bilirubin and sulfobromophthalein by isolated perfused rat liver. Gastroeneterology 83: 1163
15. Meijer DKF, Vonk RJ, Keulemans K, et al (1977) Hepatic uptake and biliary excretion of dibromosulphthalein. Albumin dependence, influence of phenobarbital and nafenopin pretreatment and the role of Y and Z protein. J Pharmacol Exp Ther 202: 8
16. Wolkoff AW, Goresky CA, Sellin J, et al (1979) Role of ligandin in transfer of bilirubin from plasma into liver. Am J Physiol 236: E638
17. Raymond GD, Galambos JT (1971) Hepatic storage and excretion of bilirubin in man. Am J Gastroenterol 55: 135
18. Wolkoff AW, Roy Chowdhury J, Arias IM (1983) Hereditary Jaundice and Disorders of Bilirubin Metabolism. In: Stanhury JB, Wijngaarden JB, Frederickson DS, Goldstein IL, Brown MS (eds) The metabolic basis of inherited disease. 5th Ed, McGraw-Hill, Inc, NY pp 1385–1420
19. Blanckaert N, Gollan J, Schmid R (1980) Mechanism of bilirubin diglucuronide formation in intact rats. Bilirubin diglucuronide formation in vivo. J Clin Invest 65: 1332
20. Sieg A, Heirwegh KPM (1988) Evidence against enzymic conversion in vitro of bilirubinmono- to bilirubindiglucuronide in preparations from Gunn rat liver (abstr). Gastroeneterol 80: 1349
21. Gordon ER, Meier PM, Goresky CA, Boyer JL (1984) Mechanism and subcellular site of bilirubin diglucuronide formation in rat liver. J Biol Chem 259: 5500
22. Vanstapel F, Blanckaert N (1987) On the binding of bilirubin and its structural analogs to hepatic microsomal UDP-glucuronyltransferase. Biochemistry 26: 6074
23. Vanstapel F, Licko V, Blanckaert N (1985) Kinetic evidence for bilirubin monoglucuronide (BMG) formation at a single substrate binding site of rat liver microsomal UDP-glucuronyltransferase (abstr.). Hepatology 5: 1010
24. Falany CN, Tephly TR (1983) Separation, purification and characterisation of three isenzymes of UDP-glucuronyltransferase from rat liver microsomes. Arch. Biochem Biophys 227: 248
25. Chowdhury RJ, Chowdhury RN, Falany CN, Tephly TR, Arias IM (1986) Isolation and characterization of multiple forms of rat liver UDP glucuronate glucuronosyltransferase. Biochemistry J 233: 827
26. Jackson R, Burchell B (1986) The full length coding sequence of rat liver UDP-glucuronyltransferase cDNA and comparison with other members of this gene family. Nucleic Acid Res 14: 779

27. Mackenzie PI (1986) Rat liver UDP-glucuronosyltransferase. Sequence and expression of a form of a cDNA encoding a phenobarbital-inducible form. J Biol Chem 261: 6119
28. Mackenzie PI (1986) Rat liver UDP-glucuronosyltransferase. cDNA sequence and expression of a form glucuronidating 3-hydroxyandrogens. J Biol Chem 261: 14112
29. Mackenzie PI (1987) Rat liver UDP-glucuronosyltransferase. Identification of cDNAs encoding two enzymes which glucuronidate testosterone, dihydrotestosterone, and β-estradiol. J Biol Chem 262: 9744
30. Iyanagi T, Haniu M, Sogawa K, Fujii-Kuriyanma Y, Watanabe S, Shively JE, Anan KF (1986) Cloning and characterization of cDNA encoding 3-methylcholanthrene inducible rat mRNA for UDP-glucuronosyltransferase. J Biol Chem 261: 15007
31. Hallinan T, de Brito AER (1981) Topology of endoplasmic reticular enzyme systems and its possible regulatory significance. In: Hormones and cell regulation, Dumont JE, Nunez J (eds) Vol 5, Elsevier North Holland, pp 73–95
32. Hayward D, Schiff D, Fedunec S, Chan G, Davis PJ, Poznansky MJ (1900) Bilirubin diffusion through lipid membranes
33. Nuwayhid N, Glaser JH, Johnson JC, Conrad HE, Hauser SC, Hirschberg CB (1986) Xylosylation and glucuronosylation reactions in rat liver Golgi apparatus and endoplasmic reticulum. J Biol Chem 261: 12936.
34. Vanstapel F, Blanckaert N (1986) Carrier-mediated translocation of intact UDP-glucose (UDPGlc) and related compounds into the lumen of rough endoplasmic reticulum (RER)-derived vesicles of rat liver (abstr.). Gastroenterology 90: 1778
35. Vanstapel F, Blanckaert N (1987) Endogenous esterification of bilirubin by liver microsomes. Evidence for an intramicrosomal pool of UDP-glucose and lumenal orientation of bilirubin UDP-glycosyltransferase. J Biol Chem 262: 4616
36. Wolkoff AW, Ketley JN, Waggoner JG et al (1978) Hepatic accumulation and intracellular binding of conjugated bilirubin. J Clin Invest 61: 142
37. Muraca M, Fevery J, Blanckaert N (1987) Relationships between serum bilirubins and production and conjugation of bilirubin: Studies in Gilbert's syndrome, Crigler-Najjar disease, hemolytic disorders, and rat models. Gastroenterol 92: 309
38. Scharschmidt BF (1986) Biliary secretion of bile pigments. In: Bile Pigments and Jaundice: Molecular, Metabolic, and Medical Aspects, Ostrow JD (ed) Marcel Dekker Inc, NY, pp 243–253
39. Lauff JJ, Kasper HE, Ambrose RT (1981) Separation of bilirubin species in serum and bile by high performance liquid chromatography. J Chromatogr 226: 391
40. Lauff JJ, Kasper MB, Wu T-W, Ambrose RT (1982) Isolation and characterization of a fraction of bilirubin in serum that is firmly bound to protein. Clin Chem 28: 629
41. Blanckaert N, D'Argenio G (1982) Presence of bilirubins linked covalently to albumin in serum of patients with cholestasis. Gastroenterology 82: 1222 (A)
42. Blanckaert N, Servaes R, Leroy P (1986) Measurement of bilirubin-protein conjugates in serum and application to human and rats sera. J Lab Clin Med 108: 77
43. Compernolle F, Van Hees GP, Blanckaert N, Heirwegh KPM (1978) Glucuronic acid conjugates of bilirubin-IXa in normal bile compared with postobstructive bile — Transformation of the 1-O-acyl glucuronides into 2-, 3-, and 4-O-acyglucuronides. Biochemistry J 171: 185
44. Blanckaert N, Compernolle F, Leroy P, Van Houtte R, Fevery J, Heirwegh KPM (1978) The fate of bilirubin-IXa glucuronide in cholestasis and during storage in vitro. Intramolecular rearrangement to positional isomers of glucuronic acid. Biochemistry J 171: 203
45. McDonagh AF, Palma LA, Trull FR, Lightner DA (1982) Phototherapy for neonatal jaundice. Configuretional isomers of bilirubin. J Am Chem Soc 104: 6865
46. Lightner DA (1982) Structure, photochemistry, and organic chemistry of bilirubin. In: Heirwegh PM, Brown SB (eds) Bilirubin, Vol I, Chemistry, CRC Press, Boca Raton, Florida
47. Lightner DA, Linnane III WP, Ahlfors CE (1983) Photooxygenation products of bilirubin in the urine of jaundiced phototherapy neonates. In: Rubaltelli FF, Jori G (eds) Neonatal Jaundice, New Trends in Phototherapy, Plenum Press New York
48. Ostrow JD, Celic L (1984) Bilirubin chemistry, ionization and solubilization by bile salts. Hepatology 4: 38S
49. Bonnett R, Davies JE, Hursthouse MB (1976) Structure of bilirubin. Nature 262: 326
50. Brodersen R (1982) Physical chemistry of bilirubin: Binding to macromolecules and membranes.

In: Heirwegh KPM, Brown SB (eds) Bilirubin, vol I, Chemistry, CRC Press, Boca Raton, Florida
51. McDonagh AF, Palma LA, Lauff JJ, Wu TW (1984) Origin of mammalian biliprotein and rearrangement of bilirubin glucuronides in vivo in the rat. J Clin Invest 74: 763
52. Jansen PLM (1981) β-Glucuronidase-resistant bilirubin glucuronide isomers in cholestatic liver disease — determination of bilirubin metabolites in serum by means of high-pressure liquid chromatography. Clin Chim Acta 110: 309
53. Wu TW (1983) Delta bilirubin: the fourth fraction of bile pigments in human serum. Isr J Chem 23: 241
54. Weiss JS, Gautam A, Lauff JJ et al (1983) The clinical importance of a protein-bound fraction of serum bilirubin in patients with hyperbilirubinemia. N Engl J Med 309: 147
55. Gautam A, Seligson H, Gordon ER, Seligson D, Boyer JL (1984) Irreversible binding of conjugated bilirubin to albumin in cholestatic rats. J Clin Invest 73: 873
56. Wu TW, Sullivan SS (1982) Biliprotein in adult icteric serum — demonstrated by extension of the alkaline methanolysis procedure. Clin Chem 28: 2398
57. Jansen FH, Billing BH (1971) The identification of monoconjugates of bilirubin in bile as amide derivatives. Biochem J 125: 917
58. Blanckaert N, Heirwegh KPM (1986) Chap 3. Analysis and preparation of bilirubins and biliverdins. In: Bilirubin, Bile Pigments, and Jaundice, Ostrow JD (ed) Marcel Dekker, NY, pp 31–79
59. Salmon M, Fenselau C, Cukier JO et al (1974) Rapid transesterification of bilirubin glucuronides in methanol. Life Sci 15: 2069
60. Blanckaert N (1980) Analysis of bilirubin and bilirubin mono- and diconjugates. Determination of their relative amounts in biological fluids. Biochem J 185: 115
61. Berk PD, Jones EA, Howe RB, Berlin NI (1980) Disorders of bilirubin metabolism. In: Bondy BK, Rosenberg L (eds): Metabolic control and disease. WB Saunders, Philadelphia
62. Brodersen R (1966) Bilirubin diglucuronide in normal human blood serum. Scand J Clin Lab Invest 18: 361
63. Werner M, Tolls RE, Hultin JV, Mellecker J (1970) Influence of sex and age on the normal range of eleven serum constituents. Z Klin Chem Klin Bioch 8: 105
64. Owens D, Evans J (1975) Population studies in Gilbert's syndrome. J Med Genet 12: 152
65. Rosenthal P, Pincus M, Fink D (1984) Sex-and age-related differences in bilirubin concentration in serum. Clin Chem 30: 1380
66. Bloomer JR, Berk PD, Howe RB, Waggoner JG, Berlin NI (1970) Comparison of fecal urobilinogen excretion with bilirubin production in normal volunteers and patients with increased bilirubin production. Clin Chim Acta 29: 463
67. Berk PD, Howe RB, Bloomer JR, Berlin NI (1969) Studies of bilirubin kinetics in normal adults. J Clin Invest 48: 2176
68. Muraca M, De Groote J, Fevery J (1983) Sex difference of hepatic conjugation of bilirubin determines its maximal biliary excretion in non-anaesthetized male and female rats. Clin Sci 64: 85
69. Muraca M, Fevery J (1984) Influence of sex and sex steroids on bilirubin uridine diphosphate-glucuronosyltransferase activity of rat liver. Gastroenterol 87: 308
70. Lester R, Schmid R (1963) Intestinal absorption of bile pigments. I. The enterohepatic circulation of bilirubin in the rat. J Clin Invest 42: 736
71. Muraca M, Fevery J, Blanckaert N, Strazzabosco M et al (1986) Different responses of serum bilirubin mono-and diconjugates to increasing bilirubin loads. J Hepatol 3: 559
72. Arias IM, Johnson L, Wolfson S (1961) Biliary excretion of injected conjugated and unconjugated bilirubin by normal and Gunn rats. Am J Physiol 200: 1091
73. Schalm L, Weber AP (1964) Jaundice with conjugated bilirubin in hyperhemolysis. Acta Med Scand 176: 549
74. Forker EL, Luxon BA (1985) Effects of unstirred Dissefluid, nonequilibrium binding, and surface-mediated dissociation on hepatic removal of albumin-bound organic anions. Am J Physiol 248: 6702
75. Van Hootegen P, Fevery J, Blanckaert N (1985) Serum bilirubins in hepatobiliary disease: Comparison with other liver function tests and changes in the post obstructive period. Hepatol 5: 112

76. Hymans van den Bergh AAH, Snapper I (1913) Deut Arch Klin Med 11: 540
77. Talafant E (1956) Properties and composition of the bile pigment giving a direct diazoreaction. Nature 178: 312
78. Billing BH, Cole PG, Lathe GH (1957) The excretion of bilirubin as a diglucuronide giving the direct Van den Bergh reaction. Biochem J 65: 774
79. Schmid R (1957) The identification of "direct-reacting" bilirubin as bilirubin glucuronide. J Biol Chem 229: 881
80. Doumas BT, Perry BW, Sasse EA, Straumfjord JV (1973) Standardization in bilirubin assays: Evaluation of selected methods and stability of bilirubin solutions. Clin Chem 19: 984
81. Jendrassik L, Grof P (1938) Vereinfachte photometrische Methoden zur Bestimmung des Blutbilirubins. Biochem Z 297: 81
82. Nosslin B (1960) The direct reaction of bile pigments in serum. Experimental and clinical studies. Scand J Clin Lab Invest 12; Suppl 49
83. Lo DH, Wu TW (1983) Assessment of the fundamental accuracy of the Jendrassik-Grof total and direct bilirubin assays. Clin Chem 29: 31
84. Tisdale WA, Kinsella ED (1959) The significance of the direct-reacting fraction of serum bilirubin in haemolytic jaundice. Am J Med 26: 214
85. Levine RA, Klatskin G (1964) Unconjugated hyperbilirubinemia in the absence of overt hemolysis, importance of acquired diseases as an etiologic factor in 366 adolescent and adult subjects. Am J Med 36: 541
86. Fevery J, Claes J, Heirwegh K, De Groote J (1967) Hyperbilirubinemia: significance of the ratio between direct-reacting and total bilirubin. Clin Chim Acta 17: 73
87. Dappen GM, Sundberg MW, Wu T-W, Babb BE, Schaeffer JR (1983) A diazo-based dry film for determination of total bilirubin in serum. Clin Chem 29: 37
88. Killenberg PG, Stevens RD, Wildermann RF, Wildermann NM (1980) The laboratory method as a variable in the interpretation of serum bilirubin fractionation. Gastroenterology 78: 1011
89. Garber GC (1981) Jendrassik-Grof analysis of total and direct bilirubin in serum with a centrifugal analyser. Clin Chem 27: 1410
90. Doumas BT, Perry BW, Jendrzejczak B, Katona V (1982) Pitfalls in the american monitor Kit methods for determination of total and "direct" bilirubin. Clin Chem 28: 2305
91. Wu T-W, Dappen GM, Spayd RW, Sundberg MW, Powers DM (1984) The Ektachem Clinical Chemistry slide for simultaneous determination of unconjugated and sugar-conjugated bilirubin. Clin Chem 30: 1304
92. Arvan D, Shirey TL (1985) Conjugated bilirubin: a better indicator of impaired hepatobiliary excretion than direct bilirubin. Ann Clin Lab Sci 15: 252
93. Weber AP, Schalm L (1962) Quantitative separation and determination of unconjugated bilirubin and conjugated bilirubin in human serum. Clin Chim Acta 7: 805
94. Heirwegh KPM, Blanckaert N (1982) Analytical chemistry of rubins. In: Bilirubin. Heirwegh KPM, Brown SB (eds) vol I, Chemistry, CRC Press, Boca Raton, Florida
95. Onishi S, Itoh S, Kawade N, Isobe K, Sugiyama S (1980) An accurate and sensitive analysis by high-pressure liquid chromatography of conjugated and unconjugated bilirubin IXa in various biological fluids. Biochem J 185: 281
96. Spivak W, Carey MC (1985) Reverse-phase HPLC separation, quantification and preparation of bilirubin and its conjugates from relative bile. Biochem J 225: 787
97. Blanckaert N, Kabra PM, Farina FA, Stafford BE et al (1980) Measurement of bilirubin and its monoconjugates and diconjugates in human serum by alkaline methanolysis and high-performance liquid chromatography. J Lab Clin Med 96: 198
98. Rosenthal P, Blanckaert N, Kabra PM, Thaler MM (1981) Liquid-chromatographic determination of bilirubin and its conjugates in rat serum and human amniotic fluid. Clin Chem 27: 1704
99. Jansen PLM, Cuypers HT, Peters WHM (1984) Quantitation of bilirubin conjugates with high-performance liquid chromatography in patients with low total serum bilirubin levels. Eur J Clin Invest 14: 295
100. Muraca M, Blanckaert N, Fevery J, Rubaltelli FF et al (1984) Serum bilirubin mono- and diconjugates in human newborns. In Rubaltelli FF, Jori G (eds): Neonatal Jaundice. Plenum Press, New York—London
101. Scharschmidt BF, Blanckaert N, Farina FA, Kabra PM et al (1982) Measurement of serum

bilirubin and its mono- and diconjugates: application to patients with hepatobiliary disease. Gut 23: 643
102. Sundberg MW, Lauff JJ, Weiss JS, Dappen GM et al (1984) Estimation of unconjugated, conjugated and "d" bilirubin fractions in serum by use of two coated thin films. Clin Chem 30: 1314
103. McKavanagh SM, Billing BH (1987) The use of Band-Elut for the estimation of serum bile pigments bonded covalently to albumin. Biomed Chromatogr 2: 62
104. Seligson D, Seligson H, Wu TW (1985) An anion-exchange chromatographic method for measuring bilirubin covalently bound to albumin. Clin Chem 31: 1317
105. Okolicsanyi L, Fevery J, Billing B, Berthelot P et al (1983) How should mild, isolated unconjugated hyperbilirubinemia be investigated? Sem Liver Dis 3: 36
106. Mesa A, Fevery J, Heirwegh KPM, De Groote J (1985) Effects of ioglycamide on the hepatic transport of bilirubin and its mono- and diconjugates in the rat. Hepatology 5: 600
107. Wulkan RW, Leijnse B (1986) Alkaline phosphatase and cholestasis. Ann Clin Biochem 23: 405
108. Hardy JB, Peeples MO (1971) Serum bilirubin levels in newborn infants. Distributions and associations with neurological abnormalities during the first year of life. Johns Hopkins Med J 128: 265
109. Bakken AF, Fog J (1967) Bilirubin conjugation in the newborn. Lancet 1: 1280
110. Jacobsen J, Brodersen R, Troll D (1967) Patterns of bilirubin conjugation in the newborn. Scand J Clin Lab Invest 120: 249
111. Jansen FH, Heirwegh KPM, Devriendt A (1969) Foetal bilirubin conjugation. Lancet 1: 702
112. Winsens A, Bratlid D (1972) Unconjugated and conjugated bilirubin in plasma from patients with erythroblastosis and neonatal hyperbilirubinemia. Acta Paediat Scand 61: 405
113. Lester R (1980) Physiologic cholestasis. Gastroenterology 78: 864
114. Barbara L, Lassari R, Roda A, Aldini R et al (1980) Serum bile acids in newborns and children. Pediatr Res 14: 1222
115. Suchy FJ, Balistrieri WF, Heubi JE, Searcy JE, Pevin RS (1981) Physiologic cholestasis: elevation of the primary serum bile acid concentrations in normal infants. Gastroenterology 80: 1037
116. Muraca M, Blanckaert N, Rubaltelli FF, Fevery J (1986) Unconjugated and conjugated bilirubin pigments during perinatal development. I. Studies on rat serum and intestine. Biol Neonate 49: 90
117. De Wolf-Peeters C, De Vos R, Desmet V (1971) Histochemical evidence of a cholestatic period in neonatal rats. Pediatr Res 5: 704

Methodological and Clinical Aspects of Bile Acid Analysis in Biological Fluids

A. Roda[1], D. Festi[2], C. Armanino[3], R. Rizzoli[2], P. Simoni[2], A. Minutello[2], E. Roda[2]

[1] Istituto di Chimica Analitica, Universita' di Messina, Italy
[2] Clinica Medica, Universita' di Bologna, Italy
[3] Istituto di Analisi e Tecnologie Farmaceutiche ed Alimentari, Universita' di Genova, Italy

The more relevant physicochemical properties of bile acids (BA) in aqueous solution are described in relation to their structure. The acquisition of such information is useful to define the possible BA species that exist in a particular biological fluid. Interaction with proteins, micelles formation, complexation and ionizzation can modulate the activity of BA in solution with consequent differences in their biological and physiological properties.

Once evaluated the BA species of the analyte in that particular biological fluid, we apply the analytical method that is suitable in terms of sensitivity, specificity to determine the BA qualitative and quantitative compositions. In this paper chemical and biological methods are reported and discussed in term of a analytical performances and applicability.

The final step is the application of these methodologies to large-scale samples in order to obtain the clinical informations required. The clinical usefulness of serum bile acid determination in the diagnosis of liver disease and bile acid malabsorption syndromes is not well established. Some aspects are still unsolved, e.g., the usefulness of postprandial serum bile acid determination in the diagnosis of mild liver diseases. The measurements of the concentrations of bile acids in bile, gastric juice, urine, and stools has a clinical relevance and applications are reported in this chapter.

Progress in Clinical Biochemistry
and Medicine, Vol. 8

1 Introduction

Recent studies of bile acids (BA) have shown that naturally occurring BA differ greatly in both their physicochemical and biological properties. Moreover, the term BA is too generic to apply to an entire class of acidic steroids, and would be more appropriate for a particular BA, i.e., cholic acid (CA), chenodeoxycholic acid (CDCA), etc.

The more relevant physiological functions, metabolism, and the interaction with proteins or membranes of BA are often so different as to be practically opposite. The main difference derives from their structure and consequently from their behaviour in aqueous solutions. Some BA self-aggregate to form micelles, others are extremely insoluble, others act as organic anions.

In the last 20 years, the use of BA such as CDCA and ursodeoxycholic acid (UDCA) for the dissolution of cholesterol gallstones prompted many studies and the metabolism of most of the physiological BA has been elucidated. These physiopathological advances have not been paralleled by an improvement in BA analysis, in particular, in the development of simple analytical methodologies.

Analytical methods must take into account the low concentration and the large number of different BA molecules (between 12 and 15) present in biological fluids and, at the same time, combine ease of use and clinical practicability.

Here we review recent advances made in the field of BA analysis, focusing on methodologies that can be applied in clinical laboratories. The clinical usefulness of BA analysis in serum, bile gastric juice and other fluids is also discussed.

2 Physicochemical Properties of Bile Acids in Aqueous Solutions

Bile acids are amphipatic compounds with distinct detergent properties [1]. The most common BA are steroids, and they possess a characteristic molecular structure as shown schematically for sodium cholate in Fig. 1. The cis fusion of the A/B rings is a peculiar feature of the common BA in higher vertebrates: as a consequence, BA are 5-β-steroid derivatives. They contain 1 to 3 hydroxyls in positions 3, 6, 7, or 12 of the steroid rings, α/β oriented which can be axial or equatorial. A short branched flexible aliphatic chain is also present at the end of the steroid nucleus: this side chain terminates in a carboxyl group (which can be free, glycine or taurine conjugated). Ionization of the carboxyl group is responsible for water solubility and micelle formation, since the protonated forms of BA have extremely low solubility [2]. Like all detergents, anions of BA (bile salts) tend to self-associate in water above a critical micellar concentration (CMC) [3].

The aggregation is step-wise and only at higher concentration are large micelles formed, i.e. containing 20–100 monomers.

The behaviour in water of such molecules is consequently pH and concentration-dependent and the following equilibria are found in solution:

$$\text{conc.} < \text{CMC} \quad \text{HA (solid)} \leftrightarrow \text{HA (aq)} \leftrightarrow H^+ + A^-$$

$$\text{conc.} > \text{CMC} \quad \text{HA (aq)} \leftrightarrow H^+ + A^-;\ nA^- \leftrightarrow (A^-)_n$$

$$mHA + nA^- \leftrightarrow (A^-)_n\,(HA)_m$$

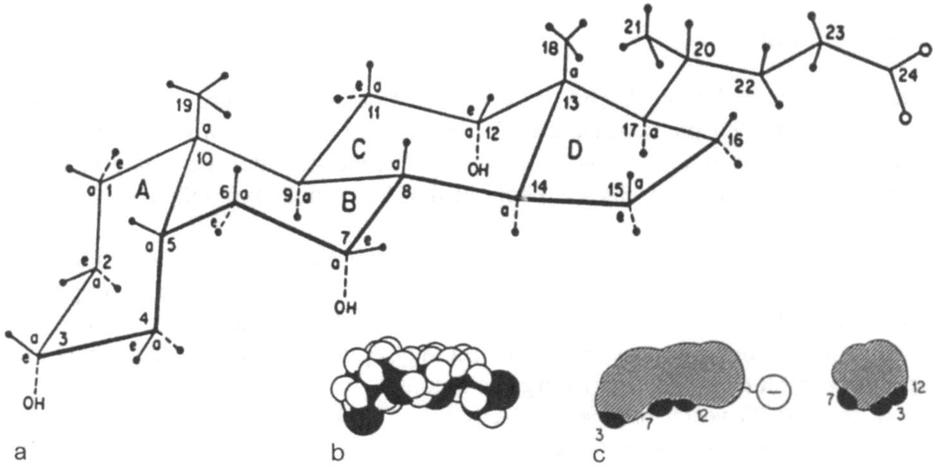

Fig. 1a. Structural formula of cholate anion, showing the cisfusion of the A/B rings and the numbering system; **b** space filling model; **c** Stuart-Briegleb molecular model in longitudinal (left) and cross sectional views (right)

where *n* is the mean aggregation number of a micelle and *m* is the mean number of undissociated BA molecules solubilized by a micelle. We previously reported[4–7] data on water solubility of the protonated form (Sw), critical micellar concentration (CMC), ionization constants and hydrophobic/hydrophilic balance, in a large series of BA.

The more relevant physicochemical properties of a representative series of free and conjugated BA are reported in Table 1.

The primary determinant of the CMC values for all the BA structures studied is the available hydrophobic surface of the bile salts which results in a back-to-back micelle formation. The larger it is, the lower is the CMC. Thus, increasing the numbers of hydroxyls increases the CMC (Fig. 2).

Among dihydroxy BA equatorial hydroxyls such as 7β or 6α, which are oriented towards the hydrophobic back of the molecule, the CMC increases (Fig. 2). The

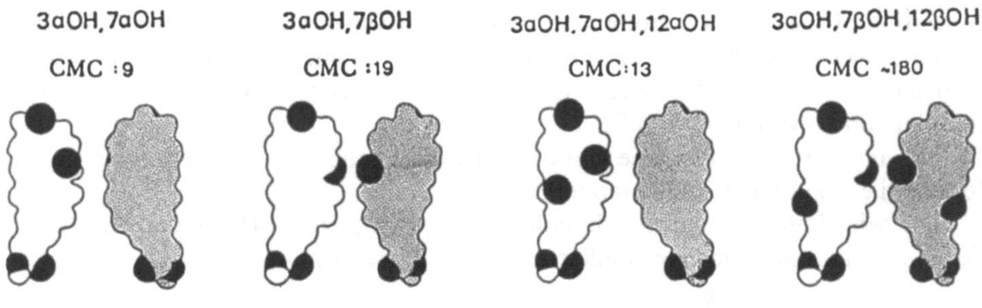

Fig. 2. Front and back (strippled) of 3,7 dihydroxy and 3,7,12 trihydroxy bile salt molecules constructed using OPK molecular models. The CMC values (mM) are reported on the top of each figure.

Table 1. Physicochemical properties of the studied bile acids

Bile Acids Trivial name	Symbol	Position and orientation of hydroxyl groups	Hydro-philicity		CMC (m*M*)		Water[a] solubility (μ*M*)
			K′	log P	water	0.15 *M* Na^+	
Cholic acid	CA	3α7α12α	1.08	0.72	13	11	273
Taurocholic acid	TCA		0.90	—	10	6	—
Glycocholic acid	GCA		0.94	0.45	12	10	32
Ursocholic acid	UCA	3α7β12α	0.26	0.20	60	39	1670
Tauroursocholic acid	TUCA		0.24	—	52	40	—
Glycoursocholic acid	GUCA		0.24	0.20	35	30	150
Hyocholic acid	HCA	3α6α7α	1.10	0.95	17	8	45
Taurohyocholic acid	THCA		—	—	—	—	—
Glycohyocholic acid	GHCA		1.00	—	—	—	—
Chenodeoxycholic acid	CDCA	3α7α	2.05	2.20	9	4	27
Taurochenodeoxycholic acid	TCDCA		1.64	—	7	3	—
Glycochenodeoxycholic acid	GCDCA		1.74	1.28	6	2	7
Ursodeoxycholic acid	UDCA	3α7β	0.95	1.05	19	7	9
Tauroursodeoxycholic	TUDCA		0.70	—	8	2.2	—
Glycoursodeoxycholic acid	GUDCA		0.72	0.75	12	4	3
Hyodeoxycholic acid	HDCA	3α6α	0.85	0.80	14	6	15
Taurohyodeoxycholic acid	THDCA		0.60	—	—	—	—
Glycohyodeoxycholic acid	GHDCA		0.75	—	—	—	—
Deoxycholic acid	DCA	3α12α	2.80	2.56	10	3	28
Taurodeoxycholic acid	TDCA		2.00	—	6	2.4	—
Glycodeoxycholic acid	GDCA		2.10	1.23	6	2	6

[a] The solubility has been measured at pH 2 for unconjugated BA and at pH 3 for glycine conjugated BA. The solubility of the protonated specie cannot be measured for tauroconjugated BA.

presence of two vicinal hydroxyls in positions 6 and 7 decreases the hydrophilicity of the shell of the hydroxyls, and is less effective than other hydroxyls (3α, 7α, 12α).

The addition of a Na^+ ion to a total concentration of 0.15 *M* lowers the CMC in a predictable manner for all BA (Table 1). This is due to a "common ion" effect of the cation which is incorporated into the aggregate, thus causing aggregation to occur at a lower concentration.

2.1 Hydrophylic/Hydrophobic Balance (HHB)

The HHB of a drug is important for its biological activity, and it is necessary to know the partition coefficient to establish a quantitative structure-activity relationship, because the partition coefficient is correlated to the transmembrane diffusion of drugs. Studies on BA partition coefficients are hampered by the peculiar behaviour of these compounds in solution. In fact, above CMC, BA anions self-aggregate and the detergent properties result in the formation of foam at the water-octanol interface. Log partition coefficients may be obtained with very low BA concentrations. Therefore, extremely sensitive methods must be used to measure the BA concentration in the two phases to obtain accurate log partition coefficients.

Since partitioning involves molecular equilibrium, the log partition coefficient represents a free energy of transfer of undissociated BA molecules from the aqueous to 1-octanol phase. Negative values suggest the undissociated BA prefers a lipidic environment.

The number and orientation of steroid hydroxyls have a great effect on the partition, while the position appears to play a minor role [8]. The change in orientation of 7 OH considerably affects the log partition coefficients, the hydroxyl being equatorially oriented towards the convex side of the molecule (β side), thus reducing the contiguous hydrophobic area.

The conjugation with glycine lowers the log partition coefficient due to the introduction of a polar amide bond.

The log partition coefficient partially agrees with the retention factors K′ obtained using reverse phase C-18 HPLC over a wide range of values.

The HPLC approach is less time-consuming than the shake-flask octanol-water method and is not subject to the same analytical problems related to the determination of BA at very low concentrations, neither is it affected by impurities in the sample. The HPLC method can be used to measure hydrophobicity only if the pH and composition of the mobile phase are standardized. In addition, since the partition process in reverse-phase chromatography is different to that in the biological phenomena, there may be discrepancies between K′ and log partition coefficient particularly when the hydroxy group is oriented toward the hydrophobic moiety of the molecule (Table 1; see CA vs UDCA).

2.2 Water Solubility

In a previous study we measured the temperature dependence of the solubility in water at pH 3.00 of a representative series of bile acids over a range between 10 and 50 °C [2]. Within each series of isomers subtle differences in both water solubility and its dependence on temperature were found. The thermodynamic functions associated with water solubility were uniformly associated with a positive enthalpy term (ΔH) and a negative entropy term (ΔS) suggesting that the dissolution process was endothermic, and was due to the solvent molecules enveloping the bulky amphiphylic molecules.

These thermodynamic function values are due to the solute-solvent interaction and the contribution derived from the physical state of the pure solid. At present, these contributions cannot be separated, and hence the effect of ΔH and ΔS values on "internal" and "external" contributions cannot be discussed. The data indicate that the solubility of BA depends largely on the hydroxy substituents. The addition of one or two hydroxy groups leads to a marked increase in water solubility. Among di- and trihydroxy BA the orientation of the hydroxy group is a key factor in water solubility. The changing of a hydroxy substituent from an alfa to a beta orientation decreases the contiguous hydrophibic area of the beta side and largely modifies the solubility.

2.3 *Ionization Properties*

In a recent study we measured the pKa of a series of BA in different ratios of a methanol/water solution. The pKa in water was estimated by extrapolating to mole fraction 0 of methanol [6–7]. The regression analysis of all pKa values gives the following equation:

$$pKa = 5.06(\pm 0.03) + 3.4(\pm 0.1)\,x$$

where x is the mole fraction of methanol, and errors are standard deviation. The correlation coefficient was $r > 0.99$ for all BA. We concluded that the pKa value of all common unconjugated BA is about 5. Thus, the ionization properties depend on the side chain of BA; the nuclear substituents, being too far from the C-24 carboxyl group, do not influence ionization.

Using the same method, we estimated the pKa of glycine conjugated BA, and found it to be in the order of 3.8 ± 0.1 in all BA studied. In this case the presence of an amide bond increases the ionization of the carboxy group by inductive effect [7].

The pKa of taurine conjugated BA has not yet been reported, but in line with other structural similarly compounds the value could be 1–2 suggesting a complete ionization of these BA at all physiological pH values.

3 Bile Acid Speciation in Biological Fluids

3.1 *Bile*

Bile is an electrolytic isotonic and isotropic solution, with a pH variable from 6.8 to 8, which contains mainly lipids (cholesterol, phospholipids, and bile salts) at a concentration of 1 to 1.5% w/w. Other important constituents are proteins (5% dry weight) and inorganic ions (Table 2).

The isotropy derives from the presence of a micellar phase, i.e. solutes are solubilized by mixed micelles composed of bile salts, cholesterol and phospholipids. An equilibrium with a vescicular phase composed of phospholipids and cholesterol occurs during bile formation in the presence of low (<CMC) concentrations of

Table 2. Composition of hepatic and gallbladder human Bile (%)

	Hepatic bile	Gallbladder bile
H_2O	97.00	85.92
Solids	2.53	14.08
Bile acids	1.93	9.14
Pigments mucins	0.53	2.98
Lipids and fatty acids	0.14	0.32
Cholesterol	0.06	0.26
Inorganic ions	0.84	0.65

BS or in the presence of a BS with low detergency (high CMC). As a consequence of the presence of many components bile is a chemically complex system in which the different species are involved in many interdependent equilibria: formation and dissolution of precipitates (insoluble lipids, i.e. cholesterol, inorganic or organic salts, etc.), acid-base equilibrium, micellization, formation of liquid crystals, complexation etc.

The metastability of this solution is consequently thermodynamically modified by the quali-quantitative composition of the above-mentioned species.

Bile acids are present at mmolar concentrations and mostly conjugated with glycine or taurine. The dilution of bile (hepatic vs gallbladder bile) the pH and the presence of counter ions affect the activity of a particular BA in a different manner. As a consequence the analytical method used takes into account the involvement of a BA anion in other equilibria (BA specific electrode, competitive binding methods), or the strong physical interaction of such molecules with lipids, proteins, etc.

3.2 Serum

The peripheral serum concentration of BA is very low when compared to that of bile (μmolar vs mmolar). Bile acids are present in serum as monomers partially bound to proteins. The major carrier protein is albumin. We recently reported microcalorimetric studies showing that the interaction is hydrophobic and is influenced by the nuclear substituents, while side-chain modifications, such as conjugation with glycine or taurine, play a minor role (Table 3) [8].

The affinity constant values of BA vs albumin are different: lithocholic acid and its conjugates are strongly albumin bound ($K_{aff} = 20 \times 10^4$ L mol^{-1}), while trihydroxy bile acids are to a lesser extent ($K_{aff} = 0.3 \times 10^4$ L mol^{-1}) [9]. Accordingly, under

Table 3. Equilibrium constant for bile acid with human serum albumin and single pass hepatic extraction in rat (range)

	k_{AFF} L/*M* 10^4	%
Cholic acid	0.33	60–66
Taurocholic acid	0.18	88–94
Glycocholic acid	0.26	68–75
Ursocholic acid	0.04	88–92
Tauroursocholic acid	—	96–98
Glicoursocholic acid	—	—
Chenodeoxycholic acid	5.5	46–52
Taurochenodeoxycholic acid	4.5	60–66
Glycochenodeoxycholic acid	4.9	56–60
Ursodeoxycholic acid	3.8	48–54
Tauroursodeoxycholic acid	—	72–80
Glycoursodeoxycholic acid	—	60–54
Deoxycholic acid	4.0	45–50
Taurodeoxycholic acid	—	55–61
Glycodeoxycholic acid	3.5	48–56

physiological conditions the unbound fraction differs: more than 40% of cholic acid is free, while lithocholic, chenodeoxycholic, and other dihydroxy BA are more than 85% albumin bound [11].

Variations in albumin or bile acid concentrations, often observed in vivo, can modify the proportion of unbound BA and consequently their liver uptake, protein-unbound bile acids can be in fact removed from peripheral blood faster and more efficiently (Table 3) [11].

Studies have shown that other proteins such as high-density lipoproteins (HDL) can bind BA and depending on BA structure, 15–40% of the bile acids can be bound to HDL.

3.3 Stools

Bile acids are present in stools at mmolar concentrations and more than 99% are in unconjugated form. The major faecal BA are secondary BA, i.e. deoxycholic acid and lithocholic acid, the latter being mainly precipitated due to the low water solubility of both protonated and ionized species. On the contrary, the amount of deoxycholic acid and other dihydroxy and trihydroxy BA in solution depends on the faecal water pH; at low pH (4.5–5) precipitation occurs, while at pH 5.5–6 these BA are ionized and consequently extremely water soluble. Protonated or ionized BA can interact with indigested foods or fibers via physical interaction thus lowering the activity of these BA in solution.

3.4 Urine

The physiological concentration of BA in urine is very low, in the order of μmolar, and the qualitative pattern is very complex. In addition to the normal glycine and taurine conjugated BA other more polar derivatives such as sulfates and glucuronides can be present particularly in liver disease patients. The pH influences the amount of these BA in solution and this only for unconjugated forms which have the highest pKa [5].

3.5 Gastric Juice

Bile acids are present in the gastric juice when a reflux of bile occurs and they can be involved in mucosal lesions through an as yet unknown mechanism. The exact definition of the BA speciation in this fluid is of extreme importance before drawing physiopathological considerations.

The low pH (2–4) limits the concentrations of theoretical BA in solution. Furthermore, when the bile reaches the stomach all unconjugated BA and glycine conjugated became protonated and thus precipitated. The only species in solution derives from the solubility product equilibrium (i.e., the solubility of the protonated species which is never higher than 0.2–0.3 m*M*) (Table 1). On the contrary, the taurine conjugated BA are always ionized and the pH does not affect the amount of tauroconjugates in solution. In conclusion, if the pH is below 3, the gastric juice contains tauroconjugates, anions and a limited amount of protonated glyco and free BA.

If the pH increases glycoconjugates become ionized and can exist in solution as anions.

Finally, at very low pH the tauroconjugate salts aggregate to form micelles at a lower concentration with respect to pH 6–7 and this means that the low pH increases the detergency of tauroconjugated BA and this in part justifies the membrane damaging effect of these BA.

4 Bile Acid Analysis

4.1 Enzymatic Methods

The first application of an enzymatic method for BA analysis was reported by Iwata and Yamasaki [12)] who used a system previously applied in steroid hormone analysis. The method involved the use of 3α hydroxysteroid dehydrogenase enzyme (3αHSD) with spectrophotometric determination of the NADH formed at 340 nm (Fig. 2). This method was not sufficiently sensitive for serum BA analysis, but adequate for the measurement of BA in bile-rich biological fluids like bile, duodenal aspirates and faeces. Now it is currently used for BA analysis in bile and in faeces and it is commercially available from various companies. Many authors have used the enzymatic method in conjunction with chromatographic procedures (TLC) which allow the separation of different BA [13,15)]. In addition, other enzymes such as 7αHSD, and 7βHSD have been isolated, purified and used in order to yield more analytical information [16–18)]. The enzymatic method acts directly on bile specimens if the concentration is higher than 1 mmol/l and a blank for each sample is included to minimize the interference of biliary pigments (bilirubin, biliverdin, etc.). 5–10 μl of bile are needed for an appropriate analysis and the method fullfils all the requisites of precision and accuracy.

If the concentration of BA is less than 1 mmol/l (duodenal aspirate) it is recommended that the BA be isolated from the matrix using reverse phase chromatography. The use of C-18 silica gel (0–18 Bondelut, Sepack 0–18) has been reported to permit a quantitative recovery of all BA [19–20)].

If information for different classes of glycine (G), taurine (T) and free (F) BA is required the enzymatic test can be preceded by a TLC separation. A good separation for G, T, or F BA may be achieved using different solvent mixtures [21)]. If different epimers i.e. UDCA, CDCA must be separated, reverse phase TLC using C-18 RF TLC plates, 0.25 mm thick, can be used. The bile or duodenal juice must be stored frozen, or at 4 °C if stored diluted 1/5 to 1/10 with 2-propanol.

The enzymatic method can be applied also to measure the BA in stools. Bile acids are largely present in stools and the extent of BA in solution depends on the pH of the faecal environment. If the pH is less than 5.5, all the unconjugated BA can precipitate as protonated acid and consequently the preanalytical step becomes critical.

The pH of the faecal sample must be immediately measured and if necessary, brought to 8–9 in order to ensure complete solubilization of all BA. If only the BA in the faecal solution is to be measured then the precipitated BA can be removed by centrifugation or filtration. The faecal sample must be diluted 1/10 w/w with 2-propanol and heated at 40 °C for 30 min. After centrifugation the supernatant is dried and reconstituted with methanol. The isopropanol in fact inhibits the enzymatic reaction.

A similar procedure can be used to evaluate the BA concentration in gastric juice. The enzymatic method can be preceded by a solid phase extraction of BA using C-18 silica gel. The necessary preliminary step is the measurement of the pH and the separation of BA eventually precipitated. Once separated, BA can be dissolved by

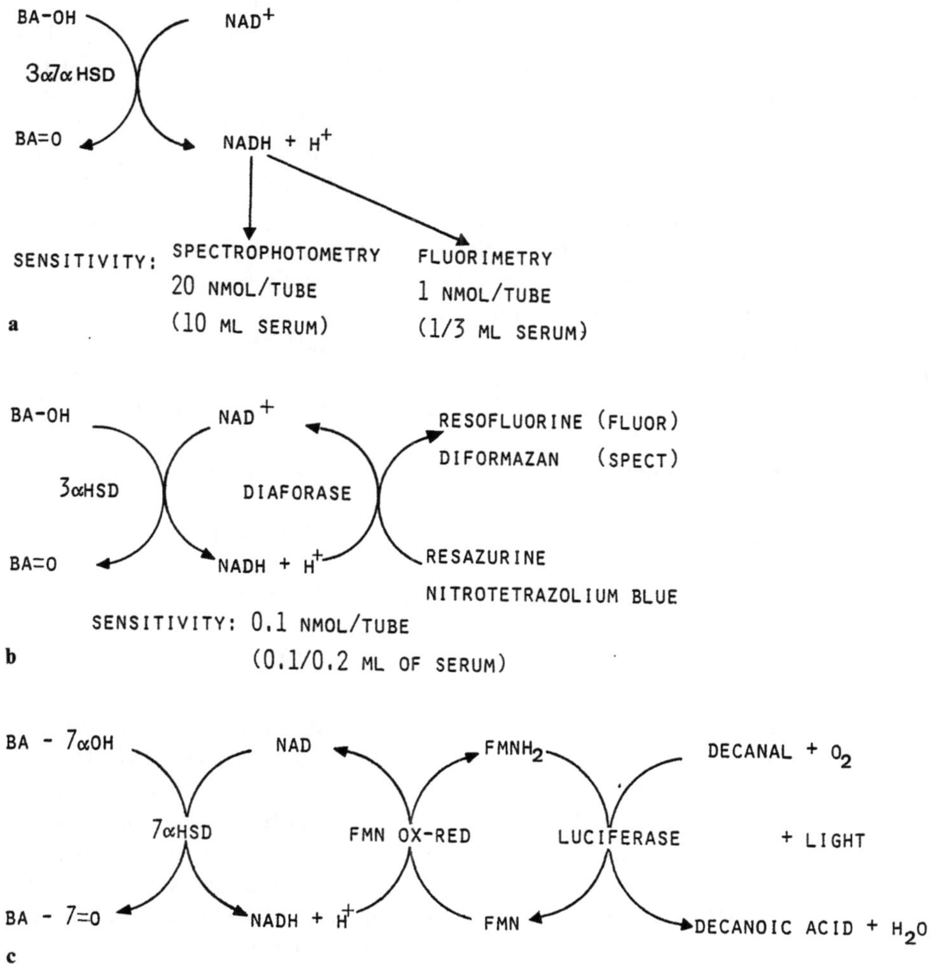

Fig. 3a–c. Principles of the enzymatic assays for bile acid: **a** enzymatic colorimetric or fluorimetric method; **b** amplified enzymatic colorimetric or fluorimetric method; **c** enzymatic bioluminescent assay

increasing the pH to 8–9 with NaOH 0.1 *M* and then applying to the C-18 column.

Many attempts have been made to apply this method to serum BA analysis. In 1970 Murphy et al.[22] reported a more sensitive (20-fold) method based on a fluorimetric determination of NADH; more recently Schwartz et al.[23] further improved the performance of the test by the use of a solid/liquid extraction of BA with Amberlite XAD 2. Despite these improvements the method is not sufficiently accurate in the determination of serum BA, or in the evaluation of the mild variations occurring in liver disease.

In 1976 Mashige et al.[24] reported an enzyme coupled system which included the use of a diaphorase in conjunction with 3αHSD. Diaphorase transfers reduction equivalents from NADH to resazurin to form fluorophore resofurin which is measured fluorimetrically at 580 nm with excitation at 560 nm. Other similar procedures were developed, but they all lacked sensitivity and clinical practicability.

An improvement was devised by Mashige et al. in 1981[25] with a coupling colorimetric procedure but the highest analytical performance was obtained by Nicolas et al.[26] with the development of an enzyme cycling method. The cycling system is composed of two enzymes: alcohol dehydrogenase and diophorase (Fig. 3). The NADH formed is oxidized by diophorase in the presence of a tetrazolium salt and then reduced again with ADH in the presence of ethanol. The final result is an accumulation of the dye which is measured colorimetrically. The authors reported good linearity and reproducibility in the range of 2 to 1000 μ*M*. Despite these significant improvements the methods were not yet entirely satisfactory for testing increased or decreased BA levels with respect to normal values.

More recently, a bioluminescent detection of NADH rendered this enzymatic technique as sensitive as the immunological methods and due to its extreme simplicity, it is suitable for large scale clinical studies and is a potential tool in clinical chemistry[27–29].

The enzymatic bioluminescent method is based on the following reactions:

$$\text{BA—OH} + \text{NAD(P)}^+ \rightarrow \text{BA=O} + \text{NAD(P)H} + \text{H}^+$$

$$\text{NADPH} + \text{FMN} + \text{H}^+ \rightarrow \text{NAD(P)}^+ + \text{FMNH}_2$$

$$\text{FMNH}_2 + \text{RCHO} + \text{O}_2 \rightarrow \text{FMN} + \text{RCOOH} + \text{H}_2\text{O} + \text{hV}$$

The first reaction is catalyzed by a specific BA hydroxysteroid dehydrogenase (3αHSD, 7αHSD, 12αHSD). The second reaction is catalyzed by NADH: FMN oxidoreductase and diophorase and the third reaction by bacterial luciferase which results in the production of light. The three enzymes can be coimmobilized on different solid supports in order to increase their stability and allow reuse for many analyses. We first developed a method based on immobilised enzymes on sepharose beads (Fig. 4a). More recently we devised[30] a continuous flow system using nylon tube (coil) with enzymes coimmobilized or separate, i.e. the specific dehydrogenase on one coil and the two bioluminescent enzymes on the other coil. This allowed us to work at two different pH and to render the method universal by changing only the specific coil (Fig. 4b).

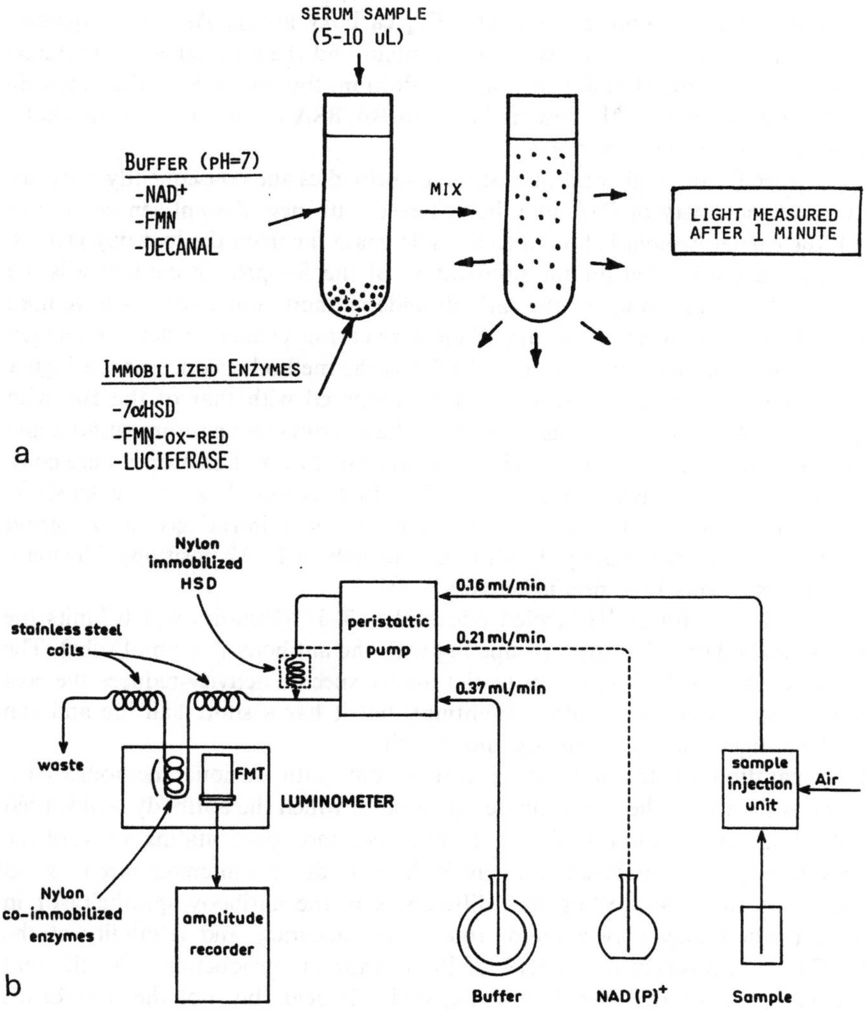

Fig. 4a, b. Schematic representation of the enzymatic bioluminescent assay for bile acid: **a** single batch assay using sepharose-immobilized enzymes; **b** continuous flow system using nylon immobilized enzymes

4.2 Immunological Methods

Immunological methods are highly sensitive and specific analytical tools for measuring analytes in biological fluids at pmolar levels and, as previously mentioned, they are suitable for the analysis of BA in serum, urine, saliva and in other biological fluids in which BA are present in very low concentrations (amniotic fluid). Since the first RIA for BA was described in 1973 by Simmonds et al.[31)], several different procedures have been reported in the literature[32–50)]. All methods utilize polyclonal antibodies produced in rabbits immunized with a bile acid coupled to a protein. The carrier protein, generally bovine serum albumin, has been covalently linked by a peptide bond on a C_{24} carboxy group. As a consequence, the side chain is completely masked by the protein and the antibodies so produced are specific only for the steroid skeleton. In addition, the low titre of the antibody produced could depend on the low molar ratio BA/BSA or on the high metabolic biotransformation of the immunogen.

A more specific and higher affinity between antibodies and consequently a greater accuracy and sensitivity of RIA may be achieved with new BA-antigen complexes in which the carrier protein is bound to the side chain far from the hydroxy groups. The method generally used for the preparation of the BA-protein conjugate is the "mixed anhydride" technique or the carbodiimide method. Some authors have used thyroglobulin instead of bovine serum albumin as carrier protein of the immunogen because of the higher molecular weight. As far as the method is concerned, a higher affinity constant of the antibody vs the BA compared with that of the BA with albumin (100–1000 times) allows us to measure the BA directly on serum without any preliminary extraction of BA or protein denaturation. Two radioisotopes were commonly utilized to label BA: ^{3}H and ^{125}I. The first isotope has been extensively used by many authors: it has the advantage of being introduced into steroid molecules and it does not modify the structure (usually at 11–12 position). The main disadvantage is its very low specific activity [31–37].

The specific activity of ^{3}H-labeled bile acid is 3–5 Ci/mmoles which limits the lowest amount of labeled antigen competing with the antibody (0.1 pmol/tube). The use of tracers such as ^{125}I with a potential higher specific activity reduces the cost of the assay (no liquid scintillation counting), but it has a short half-life and can modify the antigen's immunogenicity and stability.

The separation of the antigen is carried out with various methods. Also interesting is the use of the "solid phase" method in which the antibody is absorbed on a solid bead. Ammonium sulfate is the most common precipitating agent of the antigen-antibody complex. When different RIA methods were intercompared a good agreement was found suggesting that differences in the antibody specificity or in analytical methodology play a major role in the accuracy and reliability of the results. The main problem of the RIA of BA is antibody production. The different methods reported determine mainly conjugated bile acids but not the free forms. Moreover, each assay is specific only for a class of BA: this causes some difficulties in comparing the results obtained by different authors. Despite high sensitivity and simplicity the RIA methods are still expensive (radioactive counting, licence and disposal), so that they are limited to a few specialized laboratories [51].

4.3 Enzyme Immunoassay

Enzyme immunoassay (EIA) is an alternative analytical procedure to radioimmunoassay that exploits the specificity resulting from the use of an antibody but does not employ a radioactive isotope. Recently, many established radioimmunoassays have been changed to enzyme immunoassays because of the increased speed of analysis and reduced costs afforded by this technique.

An enzyme immunoassay has also been developed for the determination of serum BA [52–55]. Two different EIAs have been reported in the literature, one based on the "heterogeneous" principle and the other on the "homogeneous" principle. The first type of assay is similar to that of RIA: a BA covalently linked with an enzyme (peroxidase, galactosidase) is used as a tracer instead of a radioisotope.

The enzymatic activity of the tracer is recorded spectrophotometrically (after antigen-antibody complex separation) by measuring specific colour producing substrates. Matern [52] developed this method for cholic acid conjugates [52] and Maeda and Ozaki [53-55] for ursodeoxycholic acid [53,54]. The sensitivity of these methods is similar to corresponding RIA as previously reported (see above). This model of enzyme immunoassay could be extended to other BA and it may be used on a large scale in medical laboratories.

Baquir et al. [55] developed an "homogeneous" enzyme immunoassay for the determination of conjugate chenodeoxycholate. The procedure does not require extraction of serum and separation of antibody-bound from free antigen, hence the term "homogeneous". It is not based on the same principle as the common competitive or non-competitive immunoassay but on the principle that when a BA-enzyme interacts with the specific antibody the enzymatic activity is drastically reduced. The "homogeneous" method is in principle, less sensitive than the "heterogeneous" one but, it is suitable for serum BA analysis; the sensitivity is around 10 times less than similar RIA.

Recently we have developed an ultrasensitive enzyme immunoassay for the determination of conjugated chenodeoxycholic and cholic acid [56].

The method is based on a competitive principle and the antibody is immobilized on plastic beads (polystyrene microtiter plates). As a "tracer" we use a horseradish-

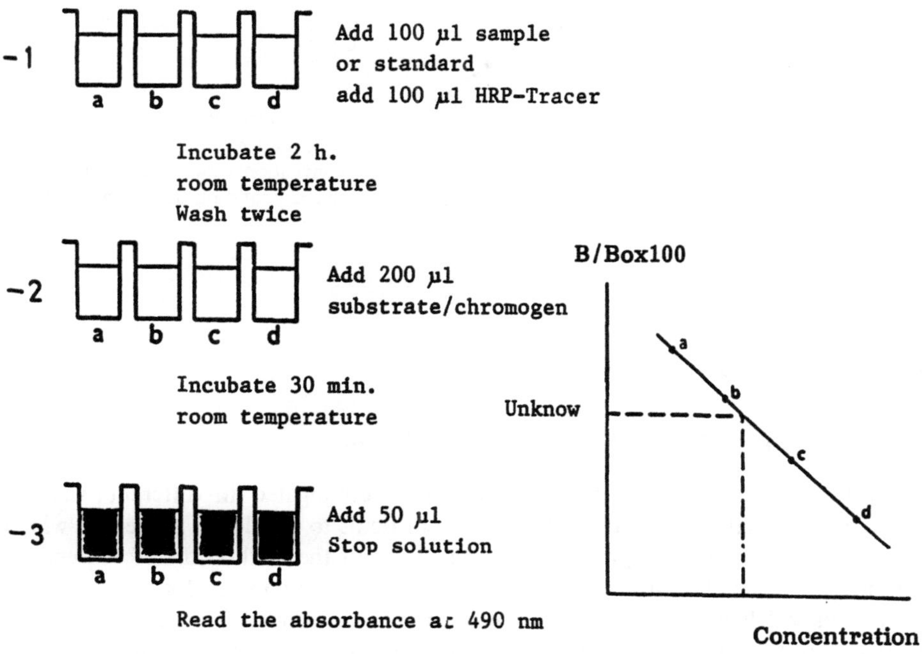

Fig. 5. Stepwise procedure for the solid-phase enzyme immunoassay for bile acids

peroxydase-BA, prepared by a slight modification of the mixed anhydride reaction. Using an appropriate amount of reagents we synthesized a "tracer" with an elevated specific activity. At a working dilution (1 μg/ml) the mass of enzyme labeled-BA is on the order of 0.1 pg. This leads to an improvement in the specific activity when compared with a titrated chenodeoxycholic acid (s.a. 3 Ci/mmol). The antibody was purified by salt precipitation; the immunoglobulin rich fraction was diluted to a final concentration of 5 μg/ml with carbonate buffer and absorbed on the plastic beads. The immobilized antibody is stable for at least 10 months if stored at 4 °C. The procedure used acts directly on serum sample (5–10 μl) and requires a 1 hour incubation of the antibody with the "tracer" and sample. After that the plastic beads are washed and the substrate (H_2O_2, *o*-phenylendiamine) is added (Fig. 5). The absorbance (490 nm) is recorded after 30 min of incubation. The sensitivity is high for EIA as shown by the mid-range of the curve (Fig. 6). In conclusion the EIA methods for BA estimation may have a sensitivity greater than the corresponding RIA due to the high activity of the tracer. In addition, this method presents the advantage of high stability of the reagent and safety for the operator.

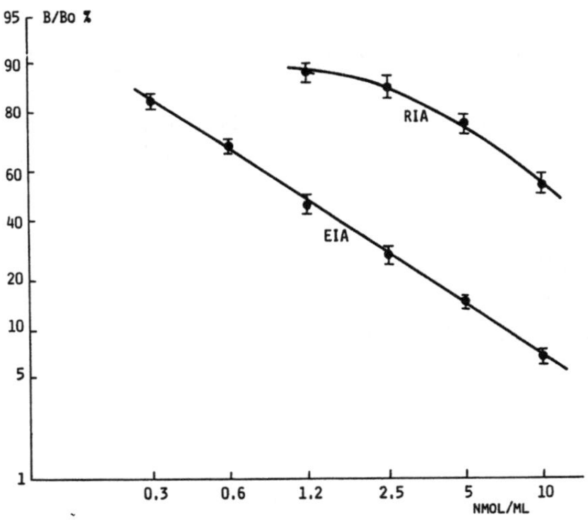

Fig. 6. Dose-response curve for chenodeoxycholic acid conjugates using a conventional radioimmunoassay and the new solid-phase enzyme immunoassay

4.4 Chromatographic Methods

4.4.1 Isolation of Bile Acid from a Biological Matrix

Since BA vary widely in polarity and can be conjugated in different ways to contain more than one negative charge, it is difficult to develop a procedure which extracts quantitatively all types of BA. The choice of the method therefore depends primarily on the nature of the biological fluid. The use of solvents for the extraction of BA was widely used but has now been essentially abandoned. The acidic properties of BA allow the use of ion-exchange chromatography for an efficient separation of the BA from neutral steroids and other lipids [57]. Anion exchanger resins such as

Amberlits (XN-1006, A-15, A-26) have been used [58-59] but the most widely employed is Amberlite XAD-2 (neutral polystyrene resin) which separates BA and their conjugates from all the biological fluids (bile, serum, urine, stools, etc.) [60-65].

In order to achieve sub-class separations such as free BA, glycine conjugates, taurine conjugates, and sulfates, gel-chromatography is more effective and Sephadex LH-20 and other derivatives of this resin such as PHP-LH-20 (piperidinohydroxy-propyl) have been successfully used (Fig. 7) [66].

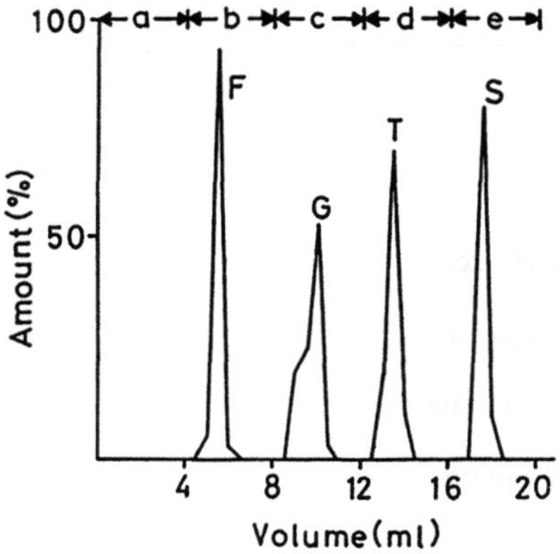

Fig. 7. Group separation of bile acids on PHP-LH-20 Eluent: a: 90% ethanol; b: 0.1 *M* acetic acid in 90% ethanol; c: 0.2 *M* formic acid in 90% ethanol; d: 0.3 *M* acetic acid — potassium acetate (pH 6.3) in 90% ethanol; e: 1% ammonium carbonate in 70% ethanol. F: free, G: glycine-conjugate, T: taurine-conjugate, S: sulfate.

More recently, reverse phase chromatography using octadecyl substituted silica (Sep-Pak C-18) has largely simplified separation from the matrix. This resin will extract both non-polar and polar BA and steroids from water and the capacity and the rate of absorption are much higher than those of polystyrene resins. Other systems have been used, particularly ion-pair/Lipidex chromatography [67,68]. Commercially available C-18 cartridges (Bond Elut, Sepack) allow the easy and rapid separation of the BA from the constituents; thus the clean-up problem, when methods like HPLC, GLC or biological methods are used, is now partially resolved [19,20,69].

4.4.2 Separation of Bile Acids

4.4.2.1 Thin-layer Chromatography

This technique is largely used for BA analysis and many solvent systems are available for the efficient separation of the common BA [70-73]. Silica gel is the absorbent most frequently used and, in order to increase the efficiency of the separation, BA derivatives can be prepared (C-24 methyl esters). Polar BA metabolites such as glucuronides and sulfates can also be separated. The detection of the separated BA is best achieved by the use of sulfuric acid, phosphomolibdates and other stains which produce fluorescent products or coloured spots [74-75].

Quantitative analysis can be performed using concentrated sulfuric acid (60–80 °C, 20 min) [76] or phosphomolybdate-cerium V ammonium sulfate followed by a direct densitometric estimation [77]. Fluorimetry can also be used with phosphoric acid and ethanol/sulfuric acid spray reagents [78-81].

Resolution of the methods can be improved using HPTLC plates or C-18 reverse phase TLC with aqueous solution as a mobile phase (methanol/water, acetonitrile/water). Using C-18 TLC it is possible for example to separate epimers or cheto BA with different hydrophobicity [82]. Table 4 shows some TLC systems used for the separation of free and conjugated BA.

Table 4. Some TLC systems for BA separation on silica-gel

Solvent system	Bile acid
Cycloexane/ethyl acetate/acetic acid (10/15/4)	unconjugates
Cycloexane/2-propanol/acetic acid (30/10/1)	unconjugates
Trimethylpentane/ethyl Acetate/acetic acid (10/10/2)	unconjugates and oxo
Isopropanol/acetic acid (93/7)	classes and conjugates
Chloroform/methanol/H_2O (70/25/3)	classes and unconjugates
Chloroform/methanol/acetic acid/H_2O (65/24/15/9)	conjugates and sulfates
Acetic acid/carbon tetrachloride/isopropilic ether/isoamile acetate/*n*-propanol/benzene (5/20/30/40/10/10)	unconjugates
Propanoic acid/isoamile acetate/*n*-propanol/H_2O (37.5/50/25/12.5)	conjugates

4.4.2.2 Gas-Chromatography

Gas-chromatography has been widely used in BA analysis and all the basic work of determining and identifying BA in different fluids has been carried out with this procedure. Various reviews have described the characteristics of these methods [83–85].

Before gas-chromatographic analysis, the BA molecules must be derived to make the BA volatile: usually the C-24 carboxy group must be esterified (Methyl ester) and eventually also the hydroxy groups (trifluoroacetates, acetates, silyl derivatives).

Many types of stationary phases have been used, e.g., non-polar methyl silicones (SE-30, OV-1), highly polar ester phases such as HiEff8BP, NGS, fluoroalkyl silicones (Q1, OV 210, SP-2401) and cyanoalkylsilicone (XE-60 OV-225) [86].

The conventional GLC methods were not totally ideal for BA analysis; in many cases there was not a good separation between epimers (UDCA vs CDCA using QF column) or between CDCA and DCA using HiEff8BP [87,88]. The analysis time is very long, in the order of 30–60 minutes, and the high number of preliminary steps (extraction from the biological matrix, hydrolysis of glycine or taurine conjugates, derivation) make this method rather unsuitable for clinical chemistry applications.

More recently the use of a capillary column has greatly improved BA resolution and shortened the analysis time. The TMS ethers of BA methyl esters can be separated on a capillary tubular column of OV-101 20-m length or 20-m column packed with barium carbonate and coated with PEG 30000 [89–90]. A typical example of a GLC chromatogram is reported in Fig. 8 [91–92].

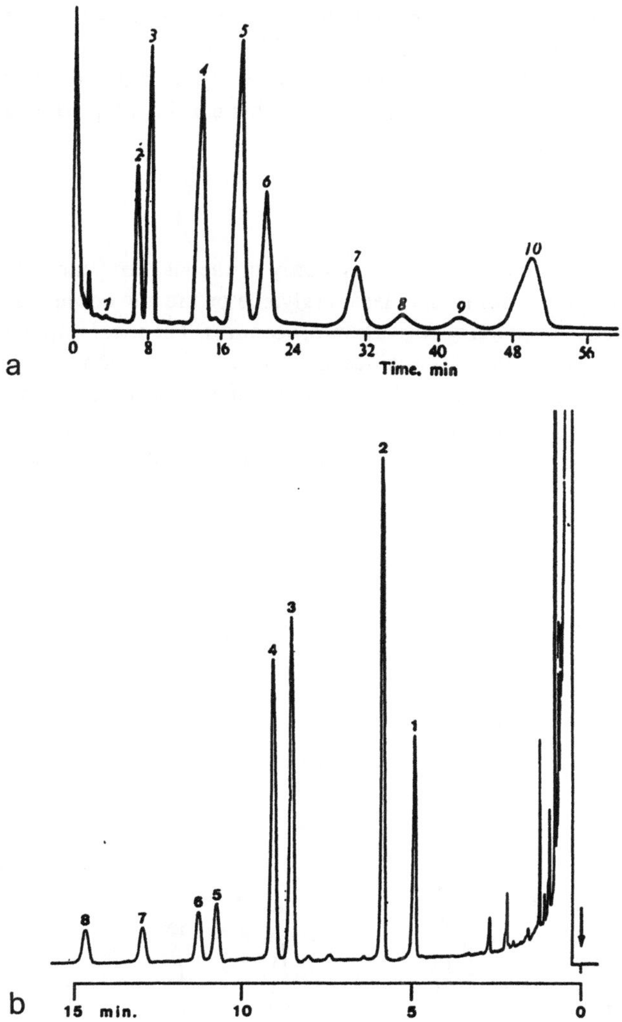

Fig. 8a, b. Typical gas-chromatographic recording of bile acid. **a** conventional GLC, methyl ester acetates, column: 180 cm × 2 mm — glass column packed with 1% OV — 225 on Gas Chrom. Q (100–120 mesh), 250 °C helium flow rate 20 ml/min. 1) Cholesterol; 2) lithocholate; 3) 3β-epimer of 2; 4) deoxycholate; 5) chenodeoxycholate; 6) hyodeoxycholate; 7) cholate; 8) hyocholate; 9) α-muricholate; 10) β-muricholate. **b** Capillary gas chromatogram of TMS ethers of methyl-esters of bile acids. A 20-m BG 20000 glass capillary column was used at 230 °C. 1) Cholesterol; 2) cholic acid; 3) deoxyholic acid; 4) chenodeoxycholic acid; 5) hyodeoxycholic acid; 6) lithocholic acid; 7) urso-deoxycholic acid; 8) 3β hydroxy-5β-cholenoic acid

The GLC is a potent analytical tool for BA analysis of different biological fluids and it is suitable for the separation and identification of a large number of hydroxy and keto BA. Quantitative analysis requires accurate standardization of the preliminary steps and the use of internal standards in order to quantify the recovery during these procedures. The most widely used detector is flame ionization and, despite its linearity with the mass injected, the detector response is different among the BA.

The coupling of GC with mass spectrometry (GC-MS) has become very popular and allows the measurement of very low concentrations of BA.

By the use of computerized GLC-MS it is possible to carry out a quantitative analysis even if there is not a perfect GLC resolution and this can be adapted as a reference method for other biological assays.

4.4.2.3 High Performance Liquid Chromatography

This technique has been widely applied in BA analysis only in the last ten years and the main advantages for GLC is that some BA classes (glycine or taurine conjugate) can be analyzed directly without preliminary derivation procedures. The most popular stationary phase used is a C-18 reverse phase silica gel and thanks to continuous increases in the column technology (controlled uniformity of the particles) many BA can be separated with high resolution [93–99]. Figure 9 shows a typical HPLC resolution of more than 12 conjugated BA is less than 40 minutes using isocratic conditions [100].

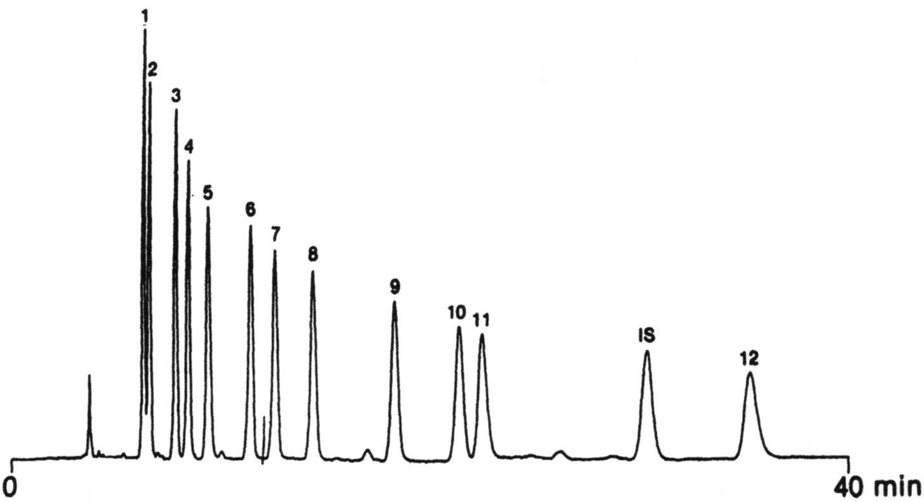

Fig. 9. HPLC chromatogram of a standard mixture of conjugated bile acid (C-18 column, isocratic elution with methanol/0.01 *M* phosphate buffer pH = 5.35 75/25 v/v and spectrophotometric detection at 200 nm). 1) Ursodeoxycholyl taurine; 2) sulfolitocholyl taurine; 3) Ursodeoxycholyl glycine; 4) cholyl taurine; 5) sulfolithocholyl glycine; 6) cholyl glycine; 7) chenodeoxycholyl taurine; 8) deoxycholyl taurine; 9) chenodeoxycholyl glycine; 10) deoxycholyl glycine; 11) lithocholyl taurine; 12) lithocholyl glycine; IS) internal standard ($3\alpha,12\alpha,5\beta$-cholanoil glycine)

The main drawback of HPLC is the detection of the separated BA; UV detection can be used below 210 nm with moderate sensitivity but cannot be used for unconjugated BA which have to be derived to form phenacyl, nitrophenacyl or *p*-bromophenacyl esters, permitting the detection of monograms of these BA [101–104] (Fig. 10).

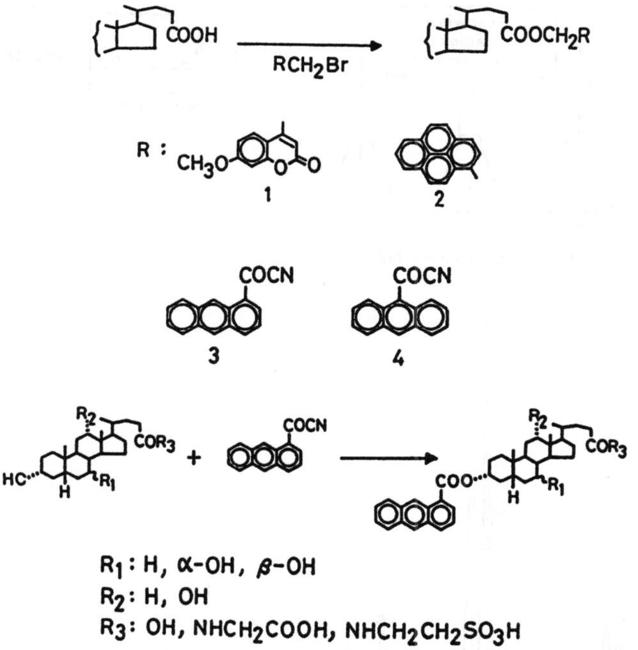

Fig. 10. Fluorescent labelling reagents for carboxy group and hydroxy group of bile acid

The reverse phase principle allows the use of aqueous methanol or acetonitrile as a mobile phase and consequently it is possible to use BA specific enzymatic detectors.

The HPLC method is well suited to BA analysis in bile due to the presence of only BA conjugated with glycine and taurine. A good resolution of all physiological BA can be easily achieved in 20–30 min including epimers (GUDCA and TUDCA). The application of this method to serum BA analysis lacks sensitivity, and the unconjugated BA largely present in serum (10–30% of total BA) cannot be analyzed directly. The clean-up procedure is not completely selective and other lipids interfere in the chromatographic profile. The alternative analytical strategy is:

1) Class separation using Sephadex LH-20 or derivatives or other resins (glycine, taurine, free acids, etc.)
2) Deconjugation of glycine and taurine conjugates (enzymatically).
3) Derivation of the free acid.

As an alternative, it is possible to use a specific and sensitive post-column detector formed by a BA specific enzyme (3α-hydroxysteroid dehydrogenase) immobilized

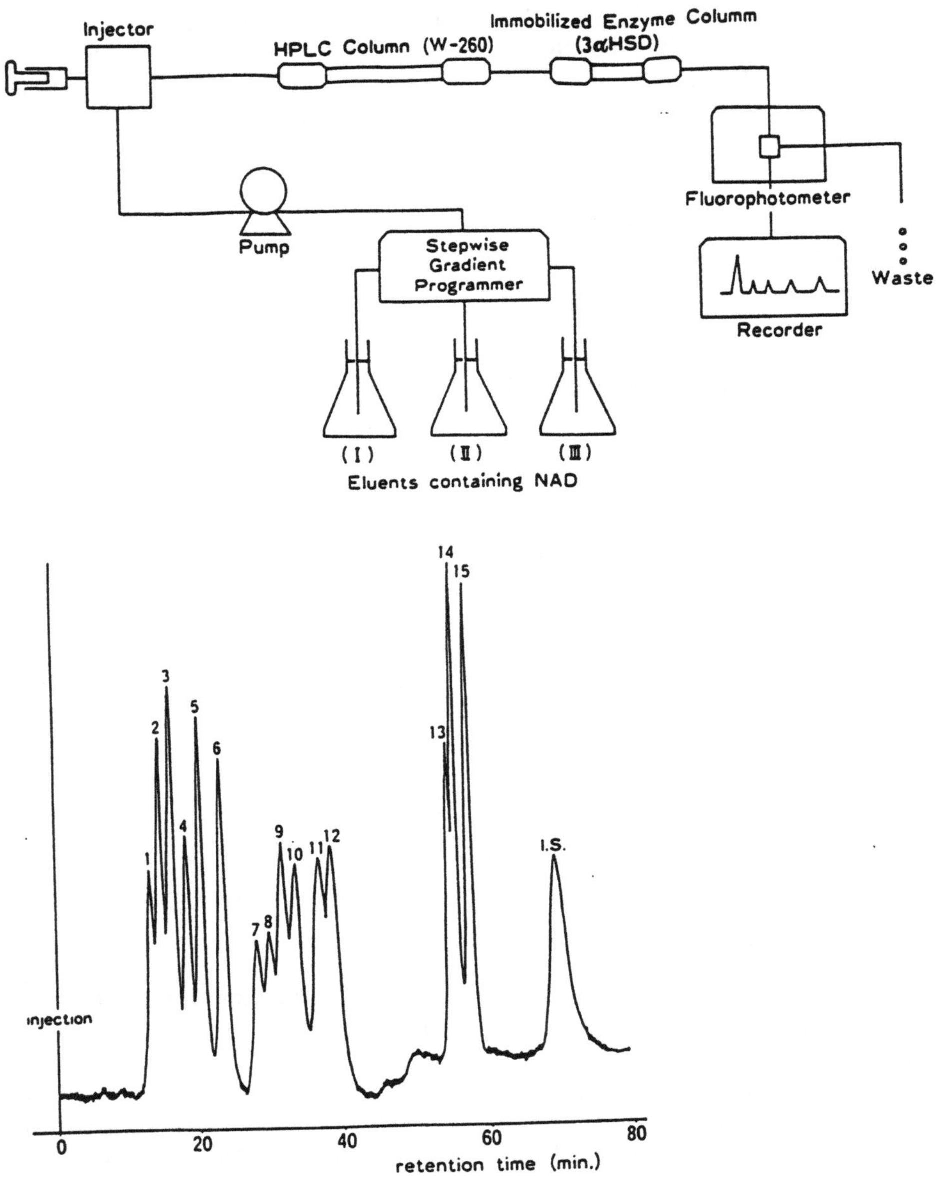

Fig. 11. Schematic representation of the HPLC system using immobilized enzyme column (3αHSD) as a detector. 1) cholic acid; 2) glycocholic acid; 3) taurocholic acid; 4) ursodeoxycholic acid; 5) glycoursodeoxycholic acid; 6) tauroursodeoxycholic acid; 7) chenodeoxycholic acid; 8) deoxycholic acid; 9) glycochenodeoxycholic acid; 10) glycodeoxycholic acid; 11) taurochenodeoxycholic acid; 12) taurodeoxycholic acid; 13) lithocholic acid; 14) glycolithocholic acid; 15) taurolithocholic acid; 16) I.S. 5-β-pregnon 3α,7α,20α triol

on a column and measuring the NADH formed at 340 nm. Using a pressure gradient more than 20 BA may be separated in a single run [105,106] (Fig. 11).

Alternative detectors are needed in BA analysis and we believe that in the near future this problem will be solved. Electrochemical detection can be used with adequate sensitivity if an hydroxy group of the BA is oxidated enzymatically to an oxo group [107–113]. Coupling of HPLC with MS is also an emerging technique. The direct identification of intact conjugated BA in biological fluids remained a problem because electron impact mass spectrometry produced spectra that either contained no molecular ion or had molecular ions of very low intensity. Using a fast atom bombardment (FAB) ion source it has been possible to obtain intense spectra of commonly occurring intact conjugated BA in isolated HPLC column fractions and in infractioned human bile [114].

5 Chemical and Biological Bile Acid Sensors

5.1 Selective Electrode for Bile Salt Anion

Surfactant ion selective electrodes for bile salts have been recently developed. The sensor is a liquid membrane containing a BA quaternary ammonium salt (esadecyltrybutilamine/octanol, benzyldimethylacetylamino cholate/decanol). A typical electrode assembly is shown schematically in Fig. 12. The electrode potential values are a direct function of the activities of the corresponding bile salt anions and follow

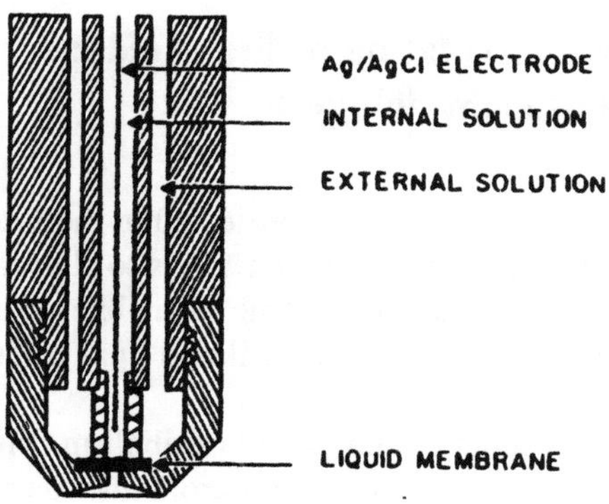

	Internal solution	External solution	
	0.01 M NaCl	$Bu_3(C_{16}H_{33})N^+$	Test Calomel
Ag/AgCl electrode	1X 10^{-4} M NaCl	Bile salts	solution
	Aqueous solution	Octanol solution	

Fig. 12. Schematic representation of a selective electrode for bile salt anion

Nernstian behaviour in premicellar BS concentration. This electrode has been successfully used to determine the bile anion-protein binding parameters [115].

5.2 *Biosensors*

Electrochemical methods based on the use of enzymes have been used for the analysis of lipids such as cholesterol [116] and phospholipids [117]. The biosensor is basically an oxidase enzyme coupled with other specific enzymes (cholesterol esterase/cholesterol oxidase, phospholipase D and choline oxidase) immobilized on a solid support (nylon) directly in front of a clark electrode which is sensitive to oxygen variations.

The electrochemical determination of BA is much more difficult and requires an enzymatic amperometric system. Specific hydroxysteroid dehydrogenase (3αHSD, 7αHSD) is coimmobilized with peroxidase on a clark electrode according to the following reactions:

$$B{-}OH + NAD^+ \xrightarrow{HSD} B{=}O + H^+ + NADH$$

$$NADH + H^+ + 1/2\ O_2 \xrightarrow[Mn^{++}]{HRP} NAD^+ + H_2O$$

The development of these supersensors will allow the determination of biliary lipid composition in real time and with the use of microelectrodes it will be possible to perform the analysis in vivo.

6 The Enterohepatic Circulation of Bile Acids: Physiology and Pathophysiology

The BA pool is dynamically contained and preserved in the enterohepatic circulation (EHC) [118] (Fig. 13). The main function of the EHC is to allow an appropriate cycling of the BA pool through the liver and the intestine where bile acids exert their important functions (to promote biliary secretion, to facilitate lipid transport and absorption, etc.) [119].

Bile acids are the end product of cholesterol catabolism (Fig. 14). The hydroxylation in the alfa position of the 7 carbon of the steroid nucleus represents the first step in the synthetic process that takes place in the liver and ends with the formation of the two primary BA: cholic acid (CA) and chenodeoxycholic acid (CDCA). Bile acids are conjugated into the liver with the aminoacids glycine and taurine and then secreted in bile to reach the intestine, where they promote fat absorption [120]. The conjugated primary BA are absorbed in the jejunum by a passive mechanism (mostly CDCA) and in the distal ileum by an efficient active transport process (mostly CA) [121,122]. Only a small amount of primary BA escapes absorption in the small intestine and reaches the colon where it is deconjugated and 7-dehydroxylated by bacterial flora to form secondary BA: deoxycholic acid (DC, from cholic acid) and

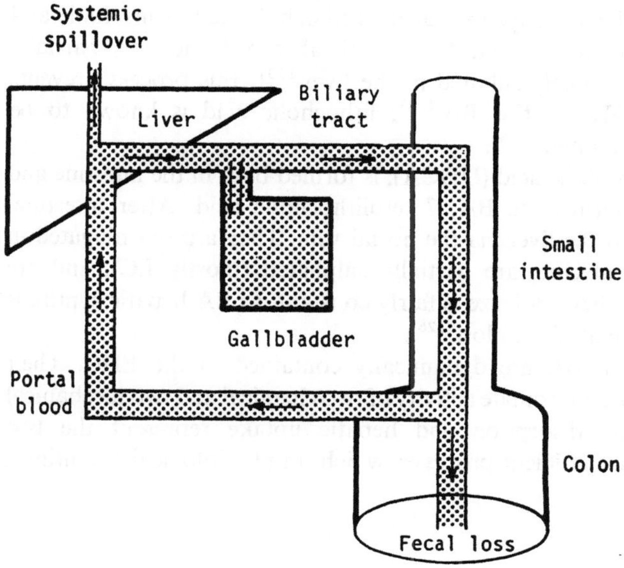

Fig. 13. The enterohepatic circulation of bile acids in humans

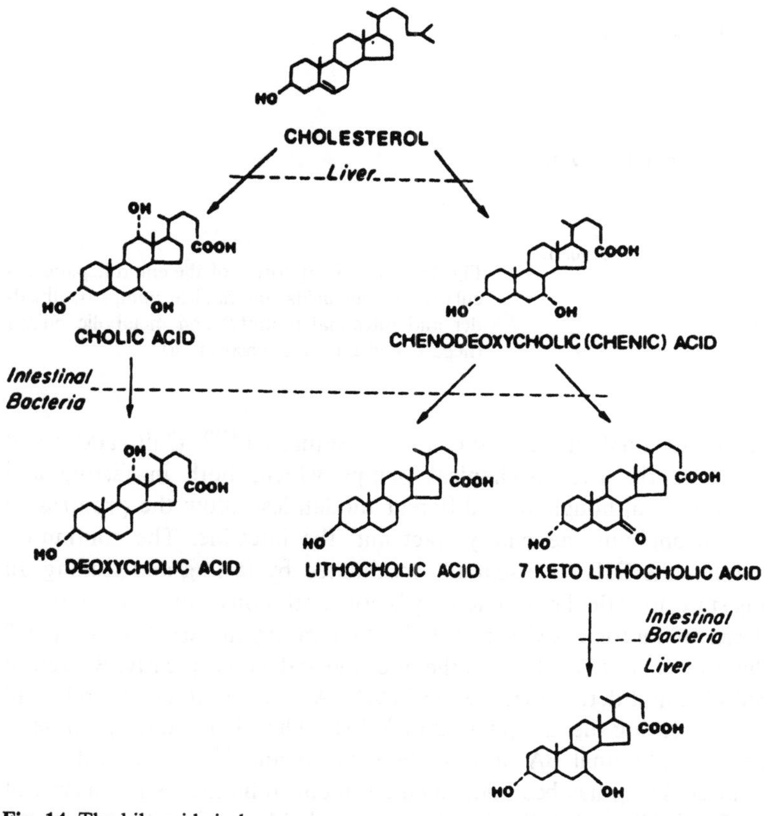

Fig. 14. The bile acids in humans

lithocholic acid (LC, from chenodeoxycholic acid). Lithocholic acid is insoluble and most of it is excreted in the stool, while DC is well absorbed and only partially excreted. Lithocholic acid is mostly sulfated in the liver [123]; this process prevents the accumulation in the EHC of this BA [124]; lithocholic acid is known to be hepatotoxic in experimental animals [125].

A tertiary BA, ursodeoxycholic acid (UDCA), is formed both in the intestine and in the liver through an intermediate BA, 7-ketolithocholic acid. After intestinal absorption, the BA return to the liver via the portal vein. They are reconjugated in the liver with taurine and glycine, are partially sulphated (mostly LC), and are again secreted into bile. The BA pool size is fairly constant, as BA hepatic synthesis in steady state conditions equals faecal loss [126].

As previously mentioned, BA are dynamically contained in the EHC. Their cycling is made possible by the coordinate activity of two chemical and two mechanical pumps (Fig. 15). Intestinal absorption and hepatic uptake represent the two chemical pumps and both are efficient processes which, in physiological conditions,

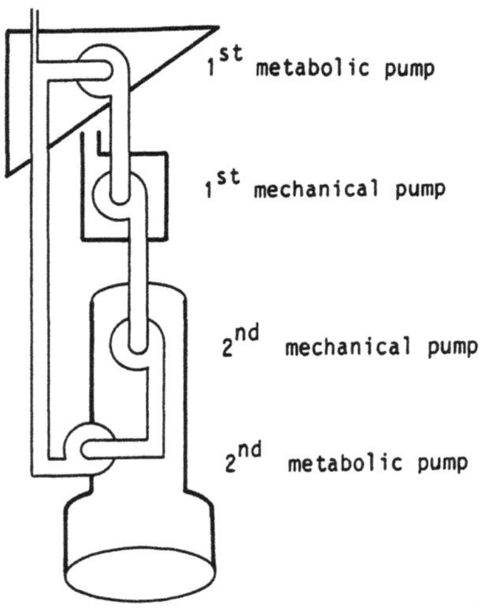

Fig. 15. The driving forces of the enterohepatic circulation of bile acids: mechanical pumps (gallbladder and intestinal motility) and metabolic pumps (hepatic and intestinal absorption)

operate below their maximal capacity and are not saturated [119]. Gallbladder and intestinal motility are the two mechanical pumps which, both in fasting and postprandial conditions, although with different modalities, allow the progression of BA molecules throughout the biliary tract and the intestine. The continuous motion of BA within the EHC is essentially modulated by fasting and feeding. In fact, in the fasting state, intestinal motor activity is not continuous, but is characterized by a cyclic pattern of motor activity called the "migrating myoelectric complex" (MMC) [127]. The transport of BA through the intestine and, consequently, BA return to the liver is pulsatile and determined by the MMC. A maximal duodenal bile acid output was observed during the late phase 2 of MMC. The MMC, during phases 2 and 3, propels the intraluminal BA pool to the distal ileum [129,130]. A gallbladder contraction (about 20–25%) has been documented in concomitance with phase 2 of the MMC [131]. After feeding, the coordination of gallbladder and gastric emptying, together with intestinal motility, produces a prolonged and steady recycling of the BA pool.

Peripheral blood, although not a part of the EHC, contains small amounts of BA and the measurements of their levels can provide an indication of the EHC dynamics [132]. After intestinal absorption, BA return, via the portal vein, to the liver where they are actively and efficiently extracted at a constant rate of about 80% [133]; the remaining BA spill over into the systemic circulation and can only be measured by a sensitive method, such as radioimmunoassay [36] (Fig. 16). Since the rate of hepatic uptake is constant and independent of the BA load [134], variations in serum BA levels will reflect intestinal input, which in turn is influenced by gallbladder and intestinal motility [132].

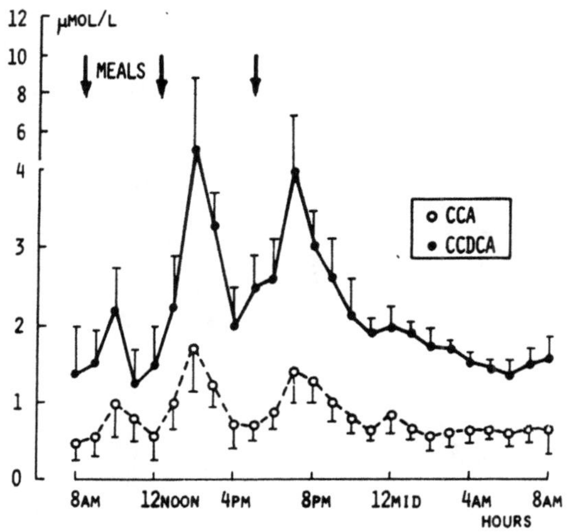

Fig. 16. Circadian rhythm of serum primary bile acids in 10 healthy subjects (mean ±SD) (↓ standard mixed meals). CCA = cholic acid conjugates; CCDCA = chenodeoxycholic acid conjugates

The EHC of BA is the result of the balance between several organs and functions. A simultaneous and complete evaluation of the EHC of BA is impossible due to its complexity and location. However, a satisfactory methodological approach to the study of the EHC dynamics is to focus on a few sampling sites (peripheral blood, duodenal bile, urine, and faeces) and to take into consideration different experimental conditions (fasting and postprandial states). In the last few years, fundamental progress has been made in the knowledge of EHC dynamics: physiological determinants and pathophysiological implications have been described and a class of diseases related to EHC disturbances has been defined and named "colanopathies" (cholesterol gallstone disease, reflux gastritis, bacterial overgrowth syndrome, primary and secondary bile acid malabsorption syndromes, ossalic urolithiasis) [135].

7 Clinical Application of Bile Acid Measurement

7.1 Bile Acids in Serum

The results of several studies [136], performed both in experimental animal systems and in man, have clearly demonstrated that serum BA concentrations are determined

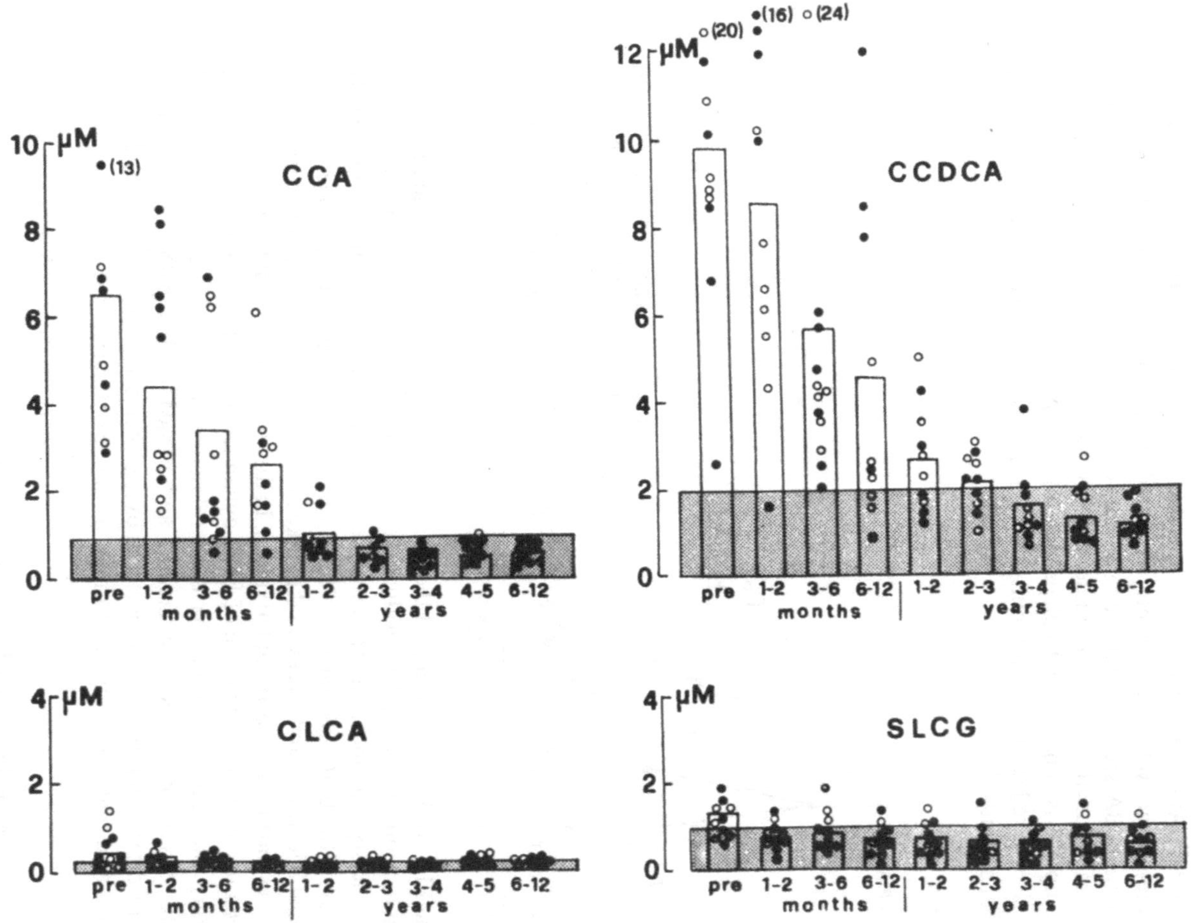

Fig. 17. Serum fasting bile acid concentrations in 90 children (45 males, 45 females) in relation to age (mean values). Shadowed area, adult normal values. CCA = cholic acid conjugates; CCDCA = chenodeoxycholic acid conjugates; CLCA = lithocholic acid conjugates; SLCG = sulfolithocholic acid conjugates. (From [141])

by the instantaneous balance between input of BA from the intestine and clearance of BA by the liver. However, since hepatic uptake, in physiological conditions, is practically constant [134-137], intestinal input is considered the main determinant of serum BA [138]. Therefore, the fraction of BA spilling into the systemic circulation is believed to be constant, i.e. independent of the load presented to the liver during eating and fasting [133]. As, in healthy subjects serum BA concentrations, and particularly their diurnal variations, are considered as a mirror of the EHC [132, 139], this should also be the case in patients with liver and intestinal diseases. We can suppose, therefore, compared to healthy subjects, serum BA will be higher in patients with liver disease and lower in patients with intestinal disease [140]. There are, however, some conditions where serum BA levels are different from normal values, but without an underlying disease. In infants, serum BA are elevated in the first year of life and reach normal values in the second year [141] (Fig. 17). This suggests a progressive maturity of the serum BA determinants throughout the first year of life, a finding which is useful in the diffential diagnosis of neonatal liver disease. In cholecystectomized subjects, serum BA levels are higher in fasting conditions, and post-prandial peaks are earlier, lower and less acute in comparison to normal subjects [142] (Fig. 18), as a consequence of the different EHC dynamics. In Gilbert's syndrome, as well as in other unconjugated hyperbilirubinemic conditions, we demonstrated [143] a lower serum level of cholic acid than in normal subjects (Fig. 19). The reported competition between bilirubin and BA may explain the faster hepatic uptake of cholic acid conjugates and, hence, their lower serum levels in these conditions.

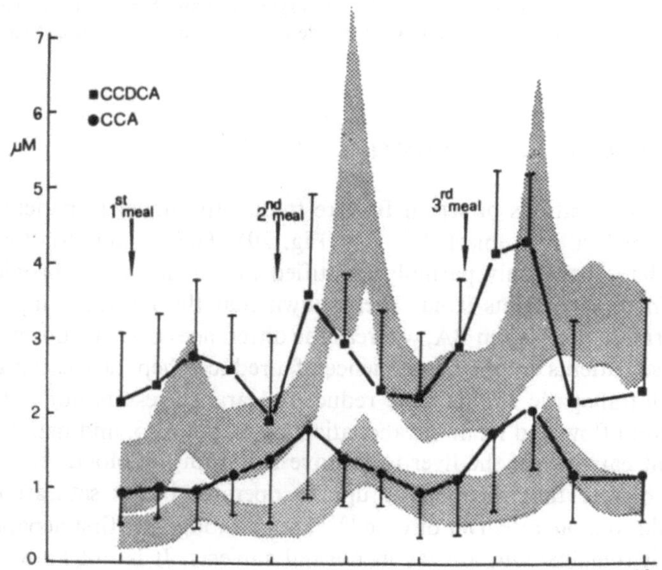

Fig. 18. Diurnal variations of serum primary bile acids in 10 cholecystectomized patients (mean ±SD). Shadowed area, normal values (↓ standard mixed meal). CCA = cholic acid conjugates; CCDCA = chenodeoxycholic acid conjugates. (From Ref. [142])

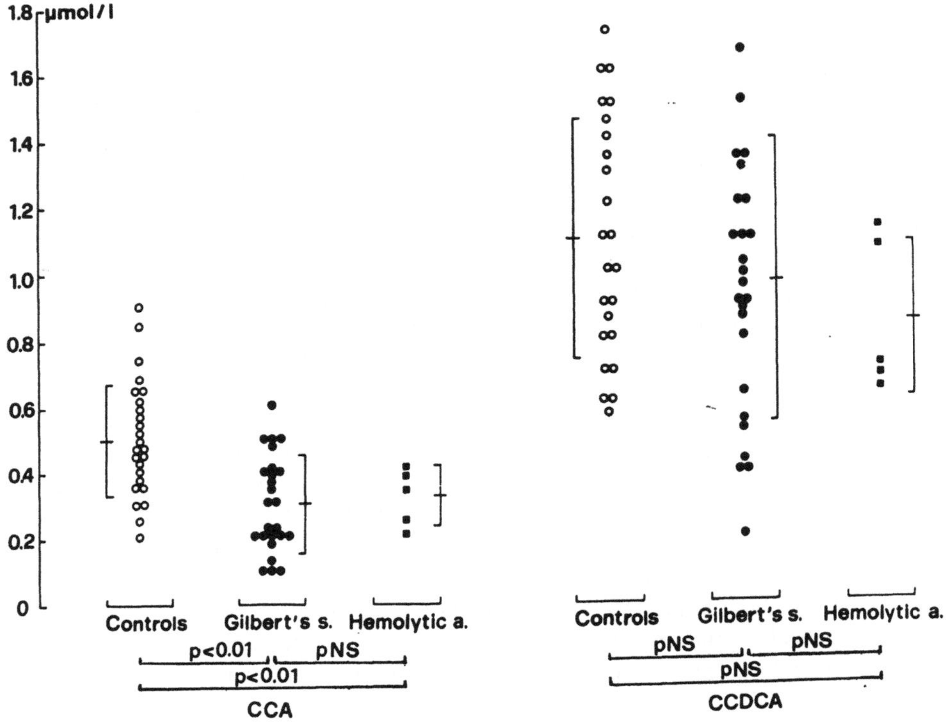

Fig. 19. Serum fasting levels of bile acids in patients with Gilbert's syndrome and hemolytic anemia, and in controls. CCA = cholic acid conjugates; CCDCA = chenodeoxycholic acid conjugates. (From [143])

7.1.1 Serum Bile Acids in Liver Disease

It is well known that concentrations of serum BA are frequently higher in patients with liver disease than in healthy subjects [144–147] (Fig. 20). The reasons for this increase in serum BA have been only partially identified in patients with chronic liver disease [148–150]. In these patients it has been shown that the intestinal input remains the major determinant of serum BA, whereas the differences between normal subjects and liver disease patients are the consequence of a reduced hepatic clearance or of the presence of intrahepatic shunts. The reduced clearance results not only from a reduced liver blood flow and from intrahepatic shunts, but also, and mainly, from a reduced inherent capacity of the liver to remove BA from the blood. However, it is important to note that the hepatic uptake operates below saturation also in patients with mild to moderate liver disease [151] and, although the first hepatic extraction is reduced, it remains constant, as in normal subjects. It is not known, however, whether the differences in serum BA between normal subjects and liver disease patients are due to differences in the first pass hepatic extraction or to a reduced V max.

In the last few years, the growth of knowledge on serum BA in health and disease has prompted studies aimed at evaluating the possibility of using serum BA in the

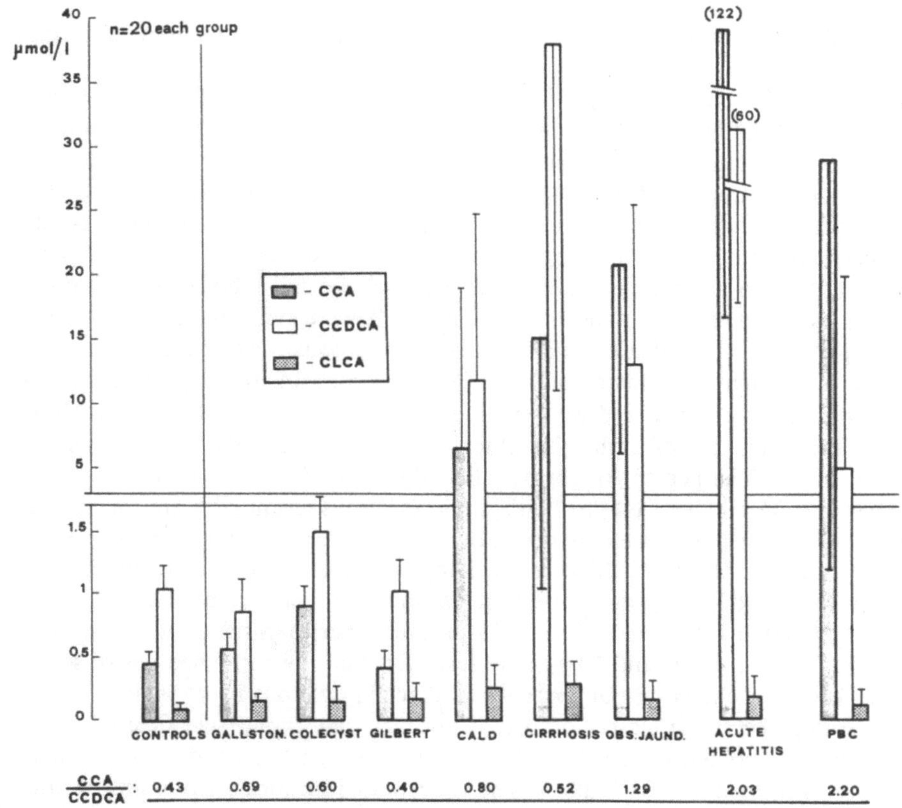

Fig. 20. Serum fasting levels of bile acids in patients with different liver diseases and controls (mean ±SD). Gallstone: gallstone patients; Colecyst: cholecystectomized patients; CALD: chronic active hepatitis; Obs. Jaund.: obstructive jaundice; PBC: primary biliary cirrhosis. CCA=cholic acid conjugates; CCDCA=chenodeoxycholic acid conjugates; CLCA=lithocholic acid conjugates. CCA/CCDCA=cholic/chenodeoxycholic acid ratio

diagnostic approach to liver disease. The main problems that have had to be overcome have been the definition of timing of the measurement (fasting or postprandial or both) and the evaluation of its accuracy.

Postprandial determination was first proposed [152)] as it was considered more sensitive than the fasting assay because it could reveal defective uptake of the endogenous BA load by a diseased liver. However, since it was later observed [151)] that in both liver disease patients and in normal subjects fractional hepatic clearance is constant in fasting and postprandial conditions, hepatic uptake in liver disease patients, although reduced, is far below its Vmax.

An alternative approach, different from fasting determination, was the administration of a known BA load, either *per os* [153)] or intravenously [154)]. However, these proposals gave rise to new problems, as the evaluation after an oral BA load is influenced by the gastric emptying and by the dissolution kinetics of the BA preparation and the evaluation after i.v. administration depends on liver blood flow [151)]. A further drawback of postprandial serum BA determination is the difficulty of standardizing sampling time. Portal-systemic shunts and alterations in intestinal and/or gallbladder motility can result in marked variability in postprandial serum BA [155].

Until recently, the definition of the diagnostic effectiveness of serum BA has been imprecise and inadequate, since most of the studies performed only compared serum BA with routine liver function tests (LFT) in terms of percent of abnormality.

To overcome these problems, we carried out a study [156] aimed at determining the sensitivity and discrimination capacity of both fasting and postprandial serum BA measurement and at comparing results with those of LFT. Fasting and postprandial (2 hrs) serum levels of cholic, chenodeoxycholic and lithocholic acid conjugates, by radioimmunoassay, and conventional LFT were evaluated in 322 patients with biopsy-proven liver disease and in 93 healthy subjects. Data were subjected to variance, discriminant and factor analyses. Fasting serum BA were higher in patients when compared to controls and were significantly higher in severe than in mild liver diseases. Determination of cholic plus lithocholic acid provided the highest discrimination capacity. The percent of correct allocation was 75.4% for LFT, 70.1% for fasting serum BA and increased to 79.6% when LFT plus serum BA were considered. Post-prandial percentages were always lower than the fasting ones. Factor analysis revealed two factors that could be related to cytolysis and protein synthesis. The serum BA was highly correlated with both factors. We therefore believe that determination of serum BA should play an adjunctive rather than substitutive role with respect to LFT, as documented by the observation that the combined use of serum BA and LFT is the most useful approach in detecting various liver diseases.

Regarding the single classes of BA, it appears that cholic acid determination has a higher sensitivity, specificity and predictive value than chenodeoxycholic acid and, when associated with LFT, allows a more accurate identification of both mild and severe disease. The significance of high serum levels of lithocholic acid requires further investigation.

Finally, the fasting serum BA determination seems to be preferable to postprandial determination because of its accuracy and ease of use.

7.1.2 Serum Bile Acids in Intestinal Disease

In the presence of an interruption of the EHC due to an intestinal disorder with or without mucosal damage, the BA are malabsorbed. Three types of BA malabsorption syndromes have been identified so far: the first, as a consequence of ileal resection or disease [157,158]; the second, considered a primary syndrome [159,160], probably related either to an increased BA synthesis, overloading the ileal transport system, or to an impaired intestinal absorption of a normal BA load due probably to a defect in the carriers of ileal mucosa, and the third type, found in uraemic [161] and cholecystectomized [142] patients. The increased amount of BA in the colon causes alterations in colonic water transport, and, consequently, diarrhoea [162]. Diagnosis of BA malabsorption [163] is based on the following procedures:

a) Measuring the fraction of ring-labelled BA (usually taurocholate) excreted in stool, 24–48 hours after i.v. administration in fasting conditions [164].

b) Calculating the excretion of 14-CO-2 in breath and of 14-C in stools after oral administration of I-14-C-glycocholic acid [165,166].

c) Evaluating serum primary BA levels throughout the day [167].

Concerning the last point, when measuring fasting and postprandial serum BA in patients with ileal resection or colectomy we observed [167] that the mean fasting levels were normal (Figs. 21, 22); after meals, serum levels of cholic acid conjugates were lower in ileal resection patients than in controls, while peaks of chenodeoxycholic acid conjugates were reduced in colectomy patients. Both serum BA postprandial peaks inversely correlated with the respective fecal excretion. In ileal resection patients we also found a significant abundance, throughout the day, of conjugates, mainly glycoconjugates, of chenodeoxycholic acid compared with cholylconjugates. These data confirm that two different EHC dynamics exist for trihydroxy and dihydroxy BA, largely determined by the intestinal absorption, and they also indicate that cholic acid postprandial peaks may be a marker of ileal disease, while those of chenodeoxycholic acid provide indirect information about colonic impairment. The evaluation, in the morning and in the evening, of the ratio between chenyl and cholylconjugates further enhances the usefulness of this test in the diagnosis of BA malabsorption.

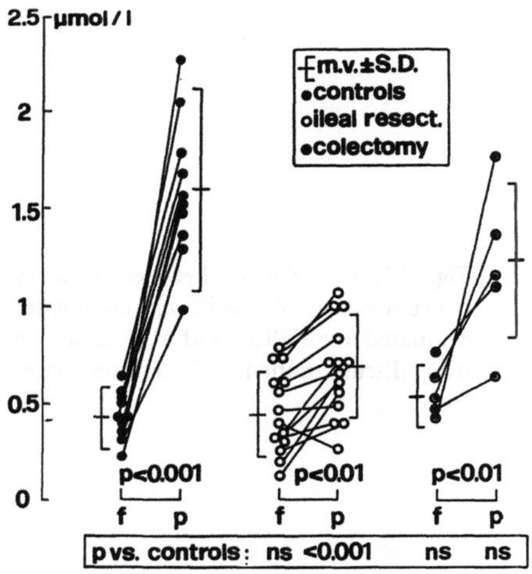

Fig. 21.

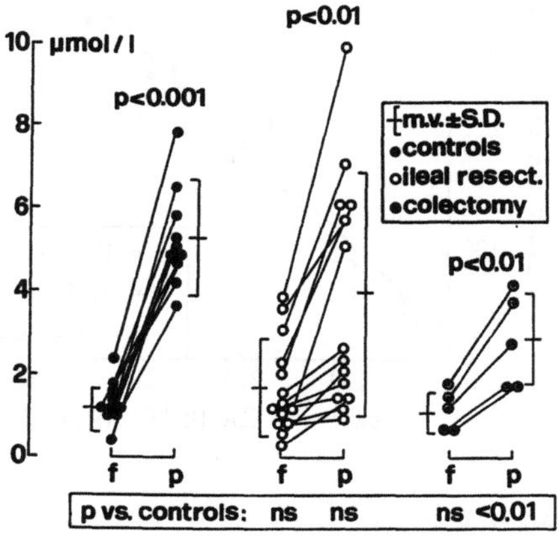

Fig. 22.

Fig. 21. Serum fasting (f) and maximal postprandial peaks (p) of cholic acid conjugates in patients with ileal resection and colectomy, and controls. (From [167])

Fig. 22. Serum fasting (f) and maximal postprandial peaks (p) of chenodeoxycholic acid conjugates in patients with ileal resection and colectomy, and controls. (From [167])

7.2 Bile Acids in Saliva

To improve further the methodology for the study of serum BA, we are developing other techniques, with the purpose of simplifying the analysis in order to propose a wider application of BA measurement. Recently we developed an enzymatic bioluminescent assay for serum BA [168], with a sensitivity comparable to that of immunological methods. Using this method we have evaluated BA concentration in saliva of patients with liver and intestinal diseases [169]. BA concentrations in saliva are independent of flow rate and reflect those in the free fraction in plasma. Conjugated cholic acid is weakly bound to albumin ($K_{aff} = 0.2 \times 10^4$ L/*M*) when compared to dihydroxy BA ($K_{aff} = 5 \times 10^4$ L/*M*) and consequently, in physiological conditions, about 40% of cholic acid conjugates are albumin unbound. Salivary BA concentrations were measured with the bioluminescent and RIA assays and with an independent method (HPLC), and the results obtained with the two methods were in good agreement ($r > 0.98$) (Fig. 23). The salivary concentrations of cholic acid conjugates were well correlated with those in serum. In control subjects the concentrations in saliva were 10–20 times lower than in serum; in patients with liver disease the salivary BA increased and the increment paralleled that of serum (Fig. 24). Furthermore, a lack of postprandial peak was observed in patients with BA malabsorption. These data demonstrate the usefulness of salivary BA determination in clinical practice. The salivary sampling has the advantage of frequent,

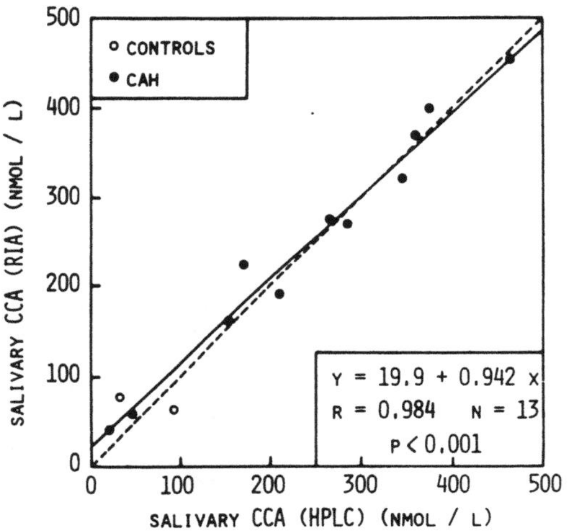

Fig. 23. Correlation between salivary concentrations of cholic acid conjugates evaluated by HPLC and RIA assay in liver disease patients (CAH) and controls.

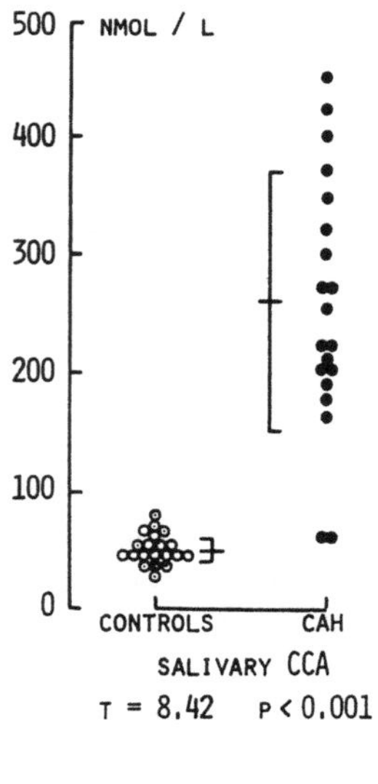

Fig. 24. Salivary levels of cholic acid conjugates in liver disease patients (CAH) and controls (mean ±SD)

easy collection by non invasive stress-free techniques. Since smaller aliquots (500 µl) can be collected at 10 or even 5 minute intervals, salivary sampling could be more useful than that of serum in short term dynamic tests.

7.3 Bile Acids in Bile

The main clinical significance of the measurement of BA in bile is related to their property of solubilization and transport of cholesterol. In bile, BA represent the catabolic product of cholesterol metabolism and serve to maintain in solution the unmetabolized cholesterol, which is an idrophobic molecule. In their role as fat solubilizer, BA are co-adjuvated by phospholipids [170]. All three together, BA, cholesterol, and phospholipids, are secreted into bile in such a proportion as to maintain bile in an isotropic phase, i.e. molecules are in solution. Disturbances in hepatic bile secretion, or interruption of EHC leading to an altered BA return to the liver, may cause secretory abnormalities Solubilizer deficiency (reduced secretion of BA and phospholipids), or excessive cholesterol secretory rate, or both, will produce an altered solubilizer/solubilizate ratio and, as a consequence, cholesterol supersaturation, crystallization, and, finally, gallstone formation [171–174].

Although the mechanisms governing BA production in the hepatic canaliculi are poorly understood, the rate of BA secretion is clearly a critical determinant of the rates of secretion of other bile components, including cholesterol, phospholipids and water. The relation between the rate of secretion of BA and that of these other substances (Fig. 25) is approximately linear at high rates of bile acid outputs, but at low rates of BA outputs, the ratio of cholesterol to BA increases and the ratio of cholesterol to phospholipids also increases.

It is possible to determine, in vitro, the limits of cholesterol solubility, in order to identify supersaturated bile from unsaturated bile. In fact, cholesterol solubility

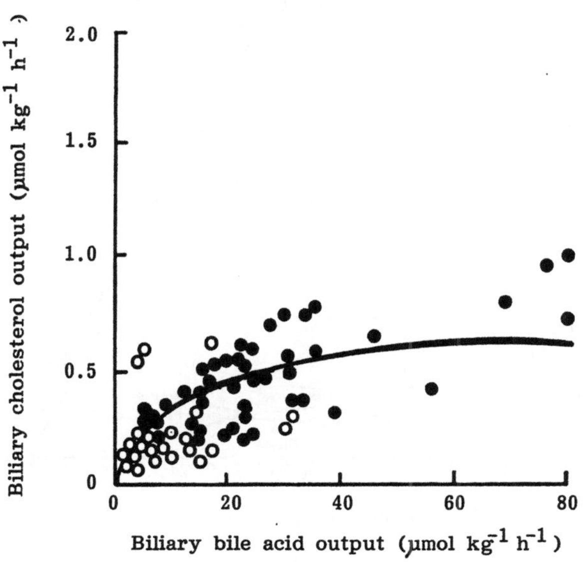

Fig. 25. Relationship between biliary cholesterol and bile acid outputs in rat. Each point represents the value for a single animal during intact (●) and (○) interrupted enterohepatic circulation.

depends on the relative concentrations of cholesterol, BA and phospholipids in bile. Admirand and Small[171] defined the upper limit of biliary cholesterol solubility from in vitro experiments in which they varied the relative concentrations of the biliary lipids. This relationship can be depicted on triangular coordinates (Fig. 26), wherein each of the three axes represent the percent of the total moles that BA, lecithin and cholesterol. The relative molar proportions of these three substances in any bile sample, therefore, are represented by a single point on the graph. Holzbach and coworkers[175] redefined the limit of cholesterol solubility and described a metastable supersaturated zone, where bile is supersaturated with cholesterol in micellar, liquid crystalline and solid crystalline phases. The liquid crystals are unstable and their cholesterol may either return to micelle or precipitate as solid crystalls. Formation of cholesterol crystals occurs slowly at the lower limit of the zone, where it may take hours or even days for cholesterol nucleation and precipitation to occur[176], and rapidly at the upper limit. Above the metastable-supersaturated zone, cholesterol also exists in the three phases. The measurement of the amount of BA, cholesterol and lecithin in a bile sample allows the calculation of the lithogenic, or saturation, index, which represents the ratio of the actual amount of cholesterol present in a bile sample to the maximum amount of cholesterol that can be dissolved in that sample. However, the observation that cholesterol in bile is solubilized and transported not only by micelles, but also by vesicles[177] and the evidence of the existence in bile of nucleating factors (either inhibiting or promoting)[178], has changed the diagnostic approach to the gallstone patient. In fact, it has been shown that the evaluation of the nucleation time better discriminates between normal and abnormal biles[179].

The composition of the BA pool represents another key factor in determining biliary cholesterol output, since the different BA have different lipid solubilization capacities, based on structural differences (number, reciprocal position and orientation

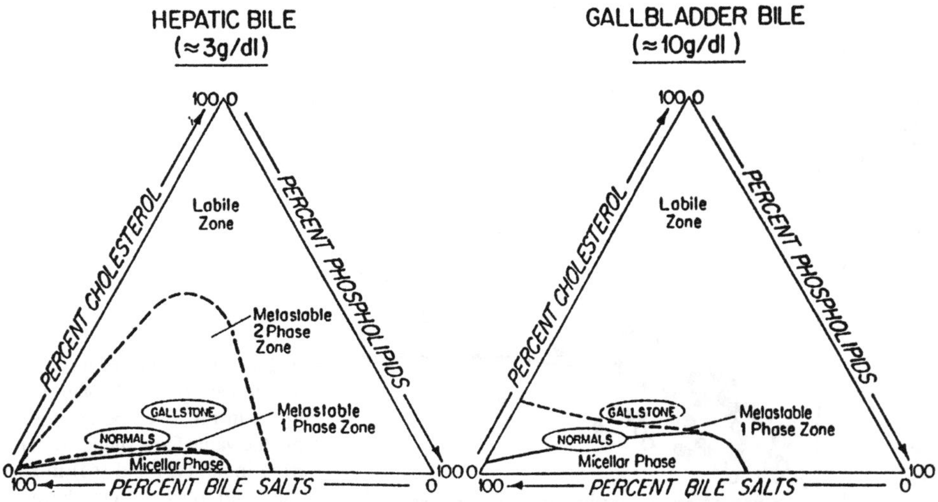

Fig. 26. Relative lipid compositions and micellar and metastable zones of human hepatic and gallbladder biles (From[119])

of the hydroxyl groups). These differences confer to BA a different hydrophilic-hydrophobic balance, which in turn influences the property of detergence, that is the capacity of the micelles to solubilize and incorporate insoluble substances, such as cholesterol [180]. From in vitro experiments [181] it has been shown that the hydrophilicity of the common bile salts, both free and conjugated, decreases in the order: ursodeoxycholic > cholic > chenodeoxycholic > deoxycholic > lithocholic acid, with taurine conjugates > glycine conjugates > free species. This hydrophilic-hydrophobic ordering has an important relationship to lecithin and cholesterol solubility since it has been shown that hydrophilicity correlates inversely with micellar dissolution rates of cholesterol monohydrate and the equilibrium cholesterol monohydrate solubilizing capacity [181]. These and other observations [182] have provided the basis for a better understanding of the mechanism of action of BA (chenodeoxycholic and ursodeoxycholic acids) in treating cholesterol gallstone patients. The enrichment of the BA pool by a given BA will change bile cholesterol secretion and saturation, depending on the changes brought about by the detergent power. This concept could be expressed by the following equation [180]:

Δ Bile cholesterol (Mol %) = Detergency of the substituted pool/detergency of the endogenous pool.

In fact, the administration of ursodeoxycholic acid, which is more hydrophilic than chenodeoxycholic acid, is associated to a fall of biliary cholesterol concentration (molar %) greater than that produced by chenodeoxycholic acid, notwithstanding the fact that the percentage of ursodeoxycholic acid in the pool never exceeds 60%

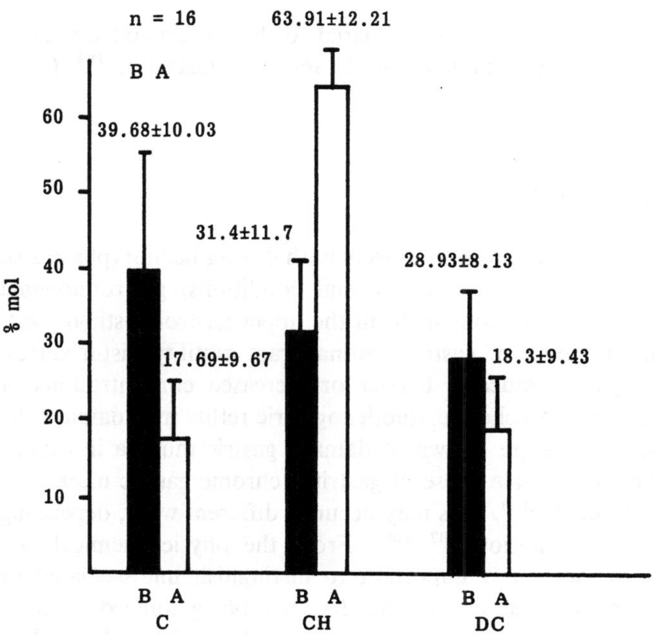

Fig. 27. Biliary bile acid pattern before (B) and during (A) chenodeoxycholic acid treatment (mean ±SD). C = cholic acid; CH = chenodeoxycholic acid; DC = deoxycholic acid. (From [184])

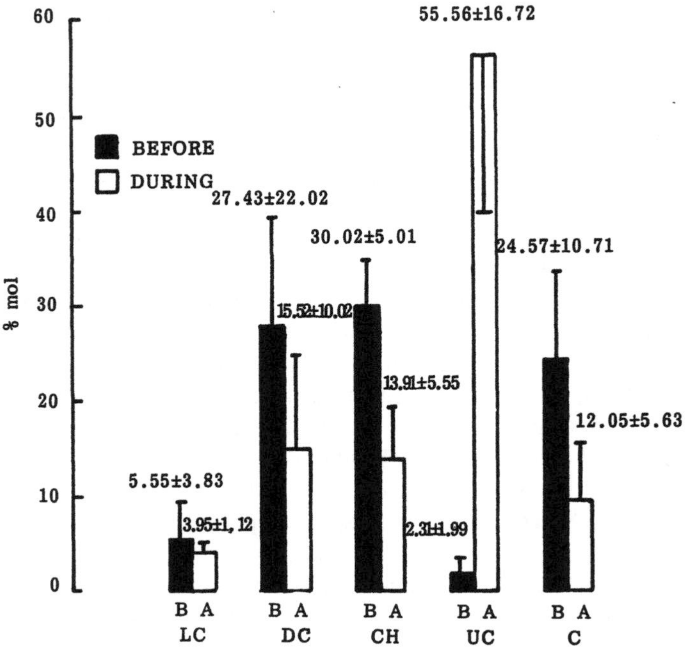

Fig. 28. Biliary bile acid pattern before (B) and during (A) ursodeoxycholic acid treatment (mean ± SD). LC=lithocholic acid; DC=deoxycholic acid; CH=chenodeoxycholic acid; US=ursodeoxycholic acid; C=cholic acid

after ursodeoxycholic acid treatment [183] as compared to that of chenodeoxycholic acid (80–90%) observed following chenodeoxycholic acid administration [184] (Figs. 27–28).

7.4 Bile Acids in Gastric Juice

Duodenogastric reflux may occur in normal subjects both during fasting (phase 2 of MMC) and postprandially [185]; however, in normal conditions, gastroduodenal motility is able to clear the intestinal content from the upper gastrointestinal tract. On the other hand, in the presence of gastrointestinal tract motility disturbances, impaired integrity of the gastric mucosal barrier or increased concentrations of hydrophobic bile salts, such as taurocholate, duodenogastric reflux may damage the gastric mucosa. In particular, BA are known to damage gastric mucosa in experimental models and are implicated as a cause of gastritis, chronic gastric ulcers and acute stress ulcerations in humans [186]. This may occur in different ways, depending on different physicochemical conditions [187–189]. From the physicochemical and pathophysiological point of view, it is important to distinguish undissociated or protonated BA from the dissociated or ionized BA, BA being ionized or non-ionized (protonated) respectively above and below their pKa's. The pKa values are the same for unconjugated BA (pKa = 5), while the pKa values are lesser than 2 for all tauroconjugated BA and the pKa of the glycoconjugated BA is 3.8.

Moreover, tauroconjugates remain in solution at a low pH, while glycoconjugates begin to precipitate out of the solution about one pH unit higher than their pKa's [3]. When protonated, BA are poorly soluble in water and differ greatly from their salts, which are water soluble above their critical micellar temperature and are step-wise self-aggregate to form micelles above a critical concentration (CMC). Ionization of the carboxyl group of the molecule is responsible for the water solubility and the micelle formation. Ionized unconjugated bile salts are appreciably less soluble than their glyco and tauroconjugates. Moreover, the number, position and orientation of the steroid hydroxyls are key determinant in BA solubilities. They are also key determinant in the CMC values, i.e. the CMC of bile salts is related to the hydrophobic/hydrophilic balance present in the molecule [3,181]. The damaging effects of BA on gastric mucosa are related to the physicochemical characteristics of both these steroids and the gastric content. The most important feature regarding mucosal damage is the influence of BA on gastric mucosal permeability. BA can disrupt the normally almost impermeable gastric mucosal barrier, with a resulting increase in transepithelial ion fluxes, H^+ from the lumen into the mucosal tissue (H^+ back diffusion) and Na^+ and K^+ from the mucosa to the lumen. Increased H^+ back diffusion, with consequent acidification of the mucosal tissue, is an important pathogenetic factor in gastric mucosal ulceration. The capacity of the BA to break this barrier is presumably attributable to their detergent properties, since it has been shown that micellar BA release phospholipids and cholesterol from the gastric mucosa [190]. For BA to exhibit these destructive effects, they must be soluble at gastric pH (taurine > glycine conjugates > free unconjugated BA); their concentration must be > 1 m*M*; hydrophobic BA are more damaging than hydrophilic BA, and the lower the pH, the more severe the resulting functional and morphologic tissue damage. On the basis of these data, a possible therapeutic approach to reflux gastritis has been proposed, i.e. the use of a hydrophilic bile acid, such as ursodeoxycholic acid. This treatment replaces gastric BA with ursodeoxycholate conjugates that have limited detergency and an inability to solubilize an appreciable amount of undissociated BA as mixed micelles; diminished gastric contents of the more hydrophobic deoxycholate, chenodeoxycholate and cholate; and it increases the glycine to taurine conjugated bile salt ratio, resulting in less total BA in solution at both neutral and acidic pH values, since precipitation pH range of glycine conjugates is 4.4–7.3 [191].

7.5 Bile Acids in Urine

In physiological conditions, BA are present in urine in negligible amounts (renal clearance < 10 μmoles/24 hrs). In patients with liver disease, particularly cholestasis, the renal excretion of BA markedly increases (> 400 μmoles/24 hrs) and the urinary BA pattern changes dramatically [119]. In particular, BA are present mainly as sulfated and glucuronidated forms. In cholestasis, the BA EHC is interrupted and a retention of BA in the liver occurs with an increase of serum BA [192], an expression of the redistribution of the BA pool (Fig. 29). Sulfation, and to a lesser extent glucuronidation, seem to play an important role in the rate of biliary, renal and intestinal BA clearance [193]. Sulfated BA are poorly absorbed by the intestine and this reduced absorption leads to an increase in faecal excretion [194]. Also BA glucuronides, present only in trace amounts in healthy subjects, have been identified in the urine in patients with cholestasis [194].

Therefore, an interruption of the EHC due to liver disease leads to an increased hydroxylation at unusual positions on the steroid nucleus, increased conjugation with taurine, glucuronidation of hydroxyl functions, all of which lead to the formation of highly polar metabolites, which are mostly eliminated by the kidney.

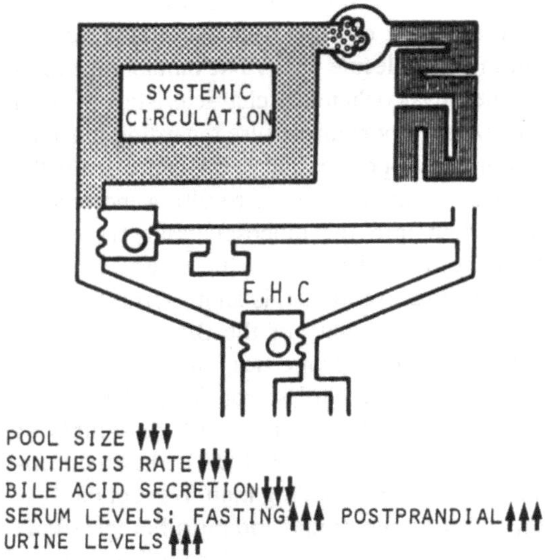

Fig. 29. Perturbations of the enterohepatic circulation (EHC) of bile acids in cholestasis (arrows indicate the effect of cholestasis on bile acid metabolism)

7.6 Bile Acids in Stools

An accurate and precise evaluation of the pattern of faecal BA is difficult, mainly because of the difficulty in separating BA from the aspecific matrix (lipids, fibers, etc.) and because of the number of the chemical forms (epimers, keto and sulfated derivatives, etc.). Most of these forms derived from BA degradation by anaerobic faecal flora [196]. Since the most common biotransformations of BA are deconjugation and 7-dehydroxylation, in healthy subjects the faecal BA are mainly composed of DCA and LCA (70–80%) and unmetabolized primary BA, CA and CDCA (7–8%) [197].

The evaluation of faecal BA excretion represents the daily synthesis of BA from cholesterol [198] and, therefore, it can be used for metabolic studies [199].

In the presence of a BA malabsorption syndrome, the faecal BA excretion increases, the faecal BA pattern changes (Fig. 30) and, as a consequence, diarrhoea may occur [198]. In fact, BA, and particularly dihydroxy BA, both free and conjugated, have cathartic properties in the colon, when present in concentrations greater than 3 m*M* in the aqueous phase of the stools [200]. Evaluating the BA faecal excretion and the dihydroxy BA concentration in the faecal water of patients with large and small ileal resections and with colectomy [201], we found that the BA malabsorption was severe in large ileal resections, mild in the small ones and slightly in colectomy (Fig. 29). Furthermore, the faecal pH proved to be a limiting factor in the occurrence of a BA diarrhoea, playing a critical role in determining the dihydroxy BA solubility in the faecal water.

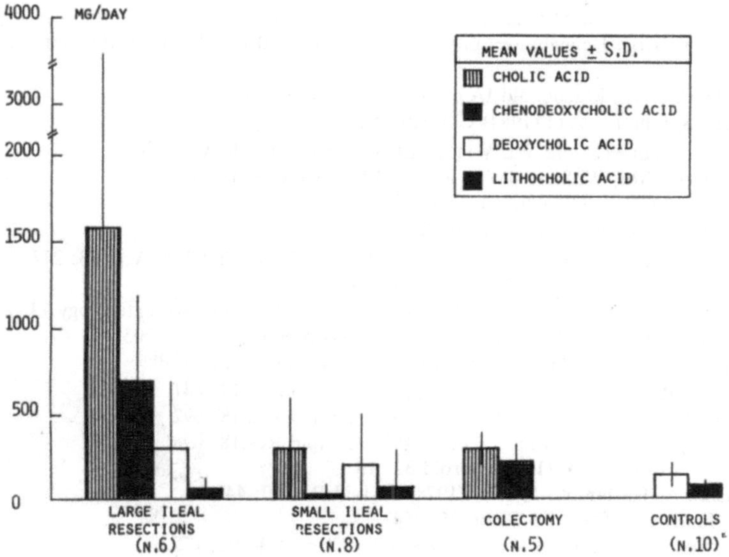

Fig. 30. Fecal bile acids in patients with large and small ileal resection, colectomy and in controls. (From [201])

8 References

1. Hofmann AF, Small DM (1967) Ann Rev Med 18: 333
2. Fini A, Roda A, Fugazza R et al (1985) J Solution Chem 14: 595
3. Small DM (1971) in: Nair PP, Kritchevsky D (eds) The Bile Acids: Chemistry, Physiology and Metabolism, vol 1, Plenum Press, New York, p 249
4. Roda A, Fini A (1984) Hepatology (Suppl) 4: 72S
5. Roda A, Hofmann AF, Mysels KJ (1983) J Biol Chem 258: 6362
6. Fini A, Roda A, De Maria P (1982) Eur J Med Chem 17: 467
7. Hofmann AF, Roda A (1984) J Lipid Res 25: 1477
8. Roda A, Fini A, Grigolo B (1988) Ann Chim 79: 1
9. Scagnolari F, Roda A, Fini A et al (1984) Biochim Biophys Acta 791: 274
10. Roda A, Cappelleri G, Aldini R et al (1982) J Lipid Res 23: 490
11. Aldini R, Roda A, Morselli AM (1982) J Lipid Res 23: 1167
12. Iwata T, Yamasaki K (1969) J Biochem 56: 424
13. Beke R, De Weerdt GA, Parijs J et al (1976) Clin Chim Acta 70: 197
14. Haslewood GAD, Murphy GM, Richardson JM (1973) Clin Sci 44: 95
15. Macdonald IA, Williams CN, Mahony DE (1974) Anal Biochem 57: 127
16. Skalhegg BA, Fasa O (1977) Scand J Gastroenterol 9: 555
17. Skalhegg BA (1974) Scand J Gastroenterol 9: 555
18. Fausa O, Skalhegg BA (1977) Scand J Gastroenterol 12: 44
19. Swhackleton CHL, Whitney JO (1980) Clin Chim Acta 107: 231
20. Whitney JO, Thaler MM (1980) J Liq Chromatogr 3: 345
21. Shepherd RW, Bunting PS, Khan M et al (1977) Clin Biochem 11: 106
22. Murphy GM, Billing BH, Baron DN (1970) J Clin Pathol 23: 594

23. Schwarz HP, Bergmann KV, Paumgartner G (1974) Clin Chim Acta 50: 197
24. Mashige F, Imai K, Osuga T (1976) Clin Chim Acta 70: 79
25. Mashige F, Tanaka N, Maki A et al (1981) Clin Chem 27: 8
26. Nicolas JC, Chaintreuil J, Descomps B et al (1980) Anal Chem 103: 170
27. Roda A, Kricka LJ, De Luca M et al (1982) J Lipid Res 23: 1354
28. Scholmerich J, van Berge Henegouwen GP, Hofmann AF et al (1984) Clin Chim Acta 137: 21
29. Sholmerich J, Hinkley JE, Macdonald IA et al (1983) Anal Biochem 133: 244
30. Roda A, Girotti S, Ghini S et al (1984) Clin Chem 30: 206
31. Simmonds WJ, Korman MG, Go WLW et al (1973) Gastroenterology 65: 705
32. Murphy GM, Edkins SM, Williams JW et al (1974) Clin Chim Acta 54: 81
33. Demers LM, Hepner GW (1976) Clin Chem 22: 602
34. Matern S, Krieger R, Gerok W (1976) Clin Chim Acta 72: 39
35. Van der Berg JWO, Van Blankenstein M, Jacobs EP et al (1976) Clin Chim Acta 73: 277
36. Roda A, Roda E, Aldini R et al (1977) Clin Chim 23: 2107
37. Schalm SW, Van Berge Henegouwen GP, Hofmann AF et al (1977) Gastroenterology 73: 285
38. Spenney JG, Hirschowitz BI, Mihas AA et al (1977) Gastroenterology 72: 305
39. Mihas AA, Spenney JS, Hirschowitz BI et al (1977) Clin Chim Acta 76: 389
40. Matern S, Krieger R, Hans C et al (1977) Scand J Gastroenterol 12: 641
41. Cowen AE, Korman MG, Hofmann AF et al (1977) J Lipid Res 18: 692
42. Cowen AE, Korman MG, Hofmann AF et al (1977) J Lipid Res 18: 698
43. Roda A, Roda E, Festi D et al (1978) Steroid 32: 13
44. Makino I, Tasiko A, Hashimoto H et al (1978) J Lipid Res 19: 443
45. Mentausta O, Janne O (1979) Clin Chem 25: 264
46. Minder E, Karlaganis G, Schmied V et al (1979) Clin Chim Acta 92: 177
47. Miller P, Weiss S, Cornell M et al (1981) Clin Chem 27: 1698
48. Minder EI (1978) J Lipid Res 20: 986
49. Baqir YA, Murison J, Ross PE et al (1979) J Clin Pathol 32: 560
50. Beckett GJ, Hunder WM, Perey-Robb IW et al (1979) Clin Chim Acta 93: 145
51. Roda A, Roda E, Aldini R et al (1980) Clin Chem 26: 1647
52. Matern S, Tietjen K, Matern A et al (1978) in: Immunoassay of hormones and drugs — European Labelled — SB PAL-Walter de Gruyter, Berlin, p 457
53. Maeda Y, Setoguchi T, Katsuki T et al (1979) J Lipid Res 20: 960
54. Ozaki A (1979) J Lipid Res 20: 2340
55. Baqir YA, Murison J, Ross PE et al (1979) Anal Biochem 93: 361
56. Roda A, Girotti S, Lodi S et al (1984) Talanta 31: 895
57. Sandberg DH, Sjovall J, Sjovall K et al (1965) J Lipid Res 6: 182
58. Alme B, Bremmelgaard A, Sjovall J et al (1977) J Lipid Res 18: 339
59. Chavez MN, Krone CL (1976) J Lipid Res 17: 545
60. Makino I, Sjovall J (1972) Anal Letters 5: 341
61. Back PZ (1976) Physiol Chem 357: 213
62. Fausa O (1975) Scand J Gastroenterol 10: 747
63. Kanno T, Tominaga K, Fujii T et al (1971) J Chromat Sci 9: 53
64. Shaw R, Elliot WH (1976) Anal Biochem 74: 273
65. Van Berge-Henegouwen GP, Hofmann AF (1976) Clin Chim Acta 73: 469
66. Goto J, Hasegawa M, Kato H et al (1978) Clin Chim Acta 87: 141
67. Hofmann AF (1967) J Lipid Res 8: 55
68. Sjovall J (1976) in: Bianchi L, Gerok W, Sickinger K (eds) Analytical methods and studies of bile acids metabolism, MTP Press, Lancaster p 67
69. Setchell KDR, Worthington J (1982) Clin Chim Acta 125: 135
70. Neher R (1969) in: Stahl E (ed) Thin layer chromatography, Springer, Berlin—Heidelberg—New York, p 351
71. Eneroth P (1969) in: Marinetti GV (ed), Lipid chromatographic analysis vol 2, Marcel Dekker, New York, p 149
72. Hofmann AF (1964) in: James AT and Morris LJ (eds), New Biochemical Separations Thin-layer chromatography of bile acids and their derivatives, London, Van Nostrand, p 283

73. Heftmann E (1976) Chromatography of steroids. Elsevier, Amsterdam, Oxford, New York, p 71
74. Arnfred T, Pedersen L (1976) Scand J Gastroent 11: 497
75. Semenuk G, Beher WT (1966) J Chromat 21: 27
76. Hara S, Tekeuchi M, Tachibana M et al (1964) Chem Pharm Bull (Tokyo) 12: 483
77. Shepherd RW, Bunting PS, Kahn M et al (1978) Clin Biochem 11: 106
78. Boucard M, Massa V, Serrano JJ et al (1974) Trav Soc Pharm Montpellier 34: 19
79. Blass KG, Persaud D (1977) Abstr Papers, Am Chem Soc Joint Conf, p 51
80. Tohma M, Nakata Y, Yamada H et al (1981) Chem Pharm Bull 29: 137
81. Taylor WA, Blass KG, Ho CS (1979) J Chromat 168: 501
82. Iida T, Yamanchi C, Chang FC (1983) J High Res Chromat 6: 617
83. Sjovall J (1964) Methods Biochem Anal 12: 97
84. Kuksis A (1966) Methods Biochem Anal 14: 325
85. Heftmann E (1976) Chromatography of steroids. Elsevier, Amsterdam, Oxford, New York, p 71
86. Yousef IM, Fisher MM, Myher JJ et al (1976) Anal Biochem 75: 538
87. Kuksis A, Gordon BA (1963) J Biochem and Physiol 41: 1355
88. Alme' B, Bremmelgaard A, Sjovall J et al (1977) J Lipid Res 18: 339
89. Laatikainen TJ, Lehtonen PJ, Hesso AE (1978) J Lab Clin Med 92: 185
90. Karlaganis G, Paumgartener G (1978) J Lipid Res 19: 771
91. Setchell KDR, Lawson AM, Blackstock EJ et al (1981) Gut 23: 637
92. Setchell KDR, Matsui A (1983) Clin Chim Acta 127: 1
93. Cantafora A, Di Biase A, Angelico M et al (1980) Clin Res 28: 628A
94. Goto J, Kato H, Nambara TJ (1980) J Liquid Chromat 3: 645
95. Maruyana K, Tanimura H, Hsaka Y (1980) Clin Chim Acta 100: 47
96. Nakayama F, Nakagaki M (1980) J Chromat 183: 287
97. Okuyama S, Uemura D, Hirata Y (1976) Chem Letters 9: 679
98. Parris NA (1977) J Chromat 133: 273
99. Shaw R, Smith JA, Elliot WH (1978) Anal Biochem 86: 450
100. Goto J, Kato H, Nambara TJ (1980) J Liquid Chromat 3: 991
101. Rossi SS, Converse JL, Hofmann AF (1987) Int J Pharm 11: 57
102. Okuyama S, Uemura D, Hirata Y (1976) Chem Letters 9: 679
103. Stellaard F, Hachey DL, Klein PD (1978) Anal Biochem 87: 359
104. Mingrone G, Greco AV, Passi S (1980) J Chromat 183: 277
105. Hasegawa S, Uenoyama R, Takeda F et al (1983) J Chromat 278: 25
106. Okuyama S, Kohubun N, Higashidate S et al (1979) Chem Letters 1443
107. Ruben AT, van Berge Henegouwen GP (1982) Clin Chim Acta 119: 41
108. Reid AD, Baker PR (1983) J Chromat 260: 115
109. Novotny M (1984) J Chromat 292: 159
110. Tietz PS, Thistle JL, Miller LJ et al (1984) J Chromat Biomed App 336: 249
111. Murana M, Goost T (1985) J Lipid Res 26: 1009
112. Reid AD, Baker PR (1985) Biochem Soc Trans 13: 1235
113. Cantafora A, Di Biase A, Alvaro D et al (1986) J Chromat (in press)
114. Whitney JO, Thalen MM, Blanckaert N et al (1983) in: Paumgartner G (ed) Bile acids and cholesterol in health and disease. MTP Press, Lancaster, p 213
115. Vadnere M, Lindebaum S (1982) Int J Pharm 11: 57
116. Mascini M, Tommassetti M, Iannello (1983) Clin Chim Acta 132: 7
117. Campanella L, Mascini G, Palleschi M et al (1985) Clin Chim Acta 151: 71
118. Hofmann AF (1977) Clin Gastroenterol 6: 3
119. Carey MC (1982) in: Arrias H, Popper H, Schacter DA, Shafritz C (eds) The liver: Biology and pathology, Raven Press, New York, p 429
120. Borgstrom B (1974) Acta Med Scand 196: 1
121. Krag E, Philips SF (1974) J Clin Invest 53: 1686
122. Dietschy JM, Small NE, Schiff ER (1972) J Clin Invest 51: 1351
123. Palmer RH (1967) Proc Natl Acad Sci 58: 1047
124. Cowen AE, Korman MG, Hofman AF et al (1975) Gastroenterology 69: 67
125. Carey JB (1964) Gastroenterology 46: 490
126. Dowling RH, Mach E, Small DM (1970) J Clin Invest 49: 232

127. Sarna SJ (1985) Gastroenterology 89: 4
128. Peeters TL, Van Trappen G, Janssen J (1980) Gastroeneterology 79: 678
129. Keane FB, Di Magno EP, Dozois RR et al (1980) Gastroenterology 78: 310
130. Scott RB, Strasberg SM, Diamant ME (1983) J Clin Invest 71: 644
131. Kraglund K, Hermind J, Jensen FT et al (1984) Scand J Gastroenterol 19: 990
132. Barbara L, Festi D, Bazzoli F et al (1985) in: Barbara L, Dowling RH, Hofmann AF, Roda E (eds) Recent advances in bile acid research. Raven Press, New York, p 29
133. Van Berge Henegouwen GP, Hofmann AF (1983) Eur J Clin Invest 13: 432
134. Reichen H, Paumgartner G (1968) Am J Physiol 231: 734
135. Van Berge Henegouwen GP, Hofmann AF (1978) Neth J Med 21: 257
136. La Russo NF, Hoffman NF, Korman MG et al (1978) Am J Dig Dis 23: 391
137. Glasinovich JC, Dumont M, Duval M et al (1975) J Clin Invest 55: 419
138. La Russo NF, Korman MG, Hoffman NF et al (1974) N Engl J Med 291: 689
139. Barbara L, Roda A, Roda E et al (1976) Rendic Gastroenterol 8: 194
140. Barbara L (1978) Ital J Gastroenterol Suppl 1: 8
141. Barbara L, Lazzari R, Roda A et al (1980) Pediatr Res 14: 1222
142. Roda E, Aldini R, Mazzella G et al (1978) Gut 19: 640
143. Roda A, Roda E, Sama C et al (1982) Gastroenterology 82: 77
144. Sherlock S, Walshe V (1948) Clin Sci 6: 223
145. Rudman D, Kendall FE (1957) J Clin Invest 36: 530
146. Carey JB (1958) J Clin Invest 37: 1494
147. Pennington CR, Ross PE, Bouchier IAD (1977) Gut 18: 903
148. Pare' P, Hoefs JC, Ashcavai M (1981) Gastroenterology 81: 959
149. Poupon RY, Poupon DE, Lebrec D et al (1981) Gastroeneterology 80: 1438
150. Ohkubo H, Okuda K, Iida S et al (1984) Gastroenterology 86: 514
151. Hofmann AF (1982) Hepatology 2: 512
152. Kaplowitz N, Kok E, Javitt NB (1973) JAMA 225: 292
153. Tashiro H (1979) Acta Hepatol Jap 20: 369
154. La Russo NF, Hoffman NE, Hofman AF et al (1970) N Engl J Med 292: 1209
155. Beckett GJ, Douglas JG, Finlayson NDC et al (1981) Digestion 22: 248
156. Festi D, Morselli Labate AM, Roda A et al (1983) Hepatology 3: 707
157. Hofmann AF, Poley JR (1972) Gastroenterology 62: 918
158. Hofmann AF (1972) Arch Intern Med 130: 597
159. Thaysen EH, Pedersen L (1976) Gut 17: 963
160. Aldini R, Roda A, Bazzoli F et al (1980) Ital J Gastroenterol 12: 251
161. Gordon SJ, Miller LJ, Kinsey MD (1976) Gut 13: 415
162. Mekhjan HS, Phillips SF, Hofmann AF (1971) J Clin Invest 50: 1569
163. Barbara L (1983) in: Barbara L, Dowling RH, Hofmann AF, Roda E (eds) Bile acids in gastroenterology, MTP Press, Lancaster, p 172
164. Stanely MM, Nemchansky B (1967) J Lab Clin Med 70: 627
165. Fromm H, Hofmann AF (1971) Lancet 2: 621
166. Roda A, Roda E, Aldini R et al (1977) Clin Chem 23: 2127
167. Aldini R, Roda A, Festi D et al (1982) Gut 23: 829
168. Girotti S, Roda A, Ghini S et al (1986) Anal Chim Acta 183: 187
169. Roda A, Roda E, Fugazza R et al (1983) Hepatology 3: 821
170. Carey MC, Small DM (1978) J Clin Invest 61: 998
171. Admirand WH, Small DM (1968) J Clin Invest 47: 1043
172. Sedaghat A, Grundy SM (1980) N Engl J Med 302: 1274
173. Bennion LJ, Grundy SM (1978) N Engl J Med 299: 1166
174. Wagner CI, Trotman BW, Soloway RD (1976) J Clin Invest 57: 473
175. Holzbach RT, Marsh M, Olszewski M et al (1973) J Clin Invest 52: 1467
176. Small DM (1980) N Engl J Med 302: 1305
177. Somjen GJ, Gilat T (1983) FEBS Lett 156: 265
178. Burnstein MJ, Ilson RG, Petrunka CN et al (1983) Gastroenterology 85: 801
179. Holzbach RT (1986) Hepatology 6: 1403
180. Carulli N, Loria P, Bertolotti M et al (1984) in: Calandra S, Carulli N, Salvioli G (eds) Liver and lipid matebolism. Excerpta Medica, Amsterdam, p 93

181. Armstrong MJ, Carey MC (1982) J Lipid Res 23: 70
182. Carulli N, Loria P, Bertolotti M et al (1983) Gastroenterology 84: 1367
183. Roda E, Roda A, Sama C et al (1979) Am J Dig Dis 24: 123
184. Barbara L, Roda E, Roda A et al (1976) Digestion 14: 209
185. Sonnenberg A, Muller-Lissner SA, Weiner HM (1982) Am J Physiol 243: 642
186. Davenport HW (1970) Gastroenterology 60: 870
187. Blanck RB, Hole D, Rhodes J (1971) Gastroenterology 61: 178
188. Eastwood GL (1975) Gastroenterology 68: 1456
189. Ritchie WP, Sheaburn EW (1976) Surgery 80: 98
190. Duane WC, Wiegard DM (1980) J Clin Invest 66: 1044
191. Stefaniwsky AB, Tint GS, Speck et al (1985) Gastroenterology 89: 1000
192. Barbara L, Roda E, Roda A et al (1979) In: Problems in intrahepatic cholestasis, Gentilini P, Popper H, Sherlock S, Teodori U (eds) Karger, Basel, p 88
193. Galeazzi R (1984) in: Calandra S, Carulli N, Salvioli G (eds) Liver and lipid metabolism, Excerpta Medica, Amsterdam, p.209
194. Makino I, Nakagawa S, Shinozaki K et al (1975) Gastroenterology 68: 545
195. Froling W, Stiehl A (1976) Eur J Clin Invest 6: 67
196. Lewis R, Gorbach S (1972) Arch Intern Med 130: 545
197. Gordbach SL, Tabaqchali S (1969) Gut 10: 963
198. Grundy SM, Ahrens EH Jr (1969) J Lipid Res 10: 92
199. Von Bergman K (1982) in: Barbara L, Dowling RH, Hofmann AF, Roda E (eds) Bile acids in gastroenterology, MTP Press, Lancaster, p 118
200. Mekhjan HS, Phillips SF, Hofmann AF (1971) J Clin Invest 50: 1567
201. Aldini R, Roda A, Festi D (1982) Dig Dis Sci 27: 495

Strategies to Integrate Laboratory Information into the Clinical Diagnosis of Hepatic and Acute Pancreatic Disease

Mario Werner

The George Washington University Medical Center, 901 Twenty-Third Street, N.W., Washington, D.C. 20037, USA

Liver diseases provide a general model for the formal analysis of the relative diagnostic weights of information obtained from different sources. Diagnosis of liver disease has three basic purposes: (1) differentiation of hemolytic, parenchymatous, and obstructive jaundice, (2) evaluation of disease course and therapeutic success, and (3) evaluation of the nature of the liver lesion. The capabilities of diagnostic findings to attain these goals can be measured according to three criteria: (1) selectivity in differential diagnosis and etiologic selectivity, (2) sensitivity, and (3) specificity with regard to morphologic changes. Applying these yardsticks to information obtained from signs and symptoms, biophysical findings and laboratory findings establishes: (1) clinical findings and liver function tests are the underpinnings of differential diagnosis, (2) liver function tests are ideal for the evaluation of the disease course, and (3) laparoscopy and liver biopsy are ideal for the evaluation of the nature of the liver lesion, but imaging procedures and liver function tests are frequently simpler means that contribute to it.

The evaluation of enzyme indicators for acute pancreatitis provides a more circumscribed model for quantitative comparison of discriminators applied to a single diagnostic entity. Convincing conclusions on what constitutes the most effective use of diagnostic enzyme indicators at present are still sparse, since valid trials of diagnostic strategies are limited by requirements for proper sampling, by restrictions on allowable observations and by judicious choice of statistical descriptors. Analogy to the three-phase system for pharmacological evaluations suggests grading criteria for diagnostic trials. Such grading of studies in the literature makes clear that some traditional enzyme indicators never have been properly validated, while the efficacy of much newer ones is relatively well-documented. A carefully executed comparison showed that urinary amylase assays are clearly inferior to serum assays of amylase, lipase, elastase, or trypsinogen which are about similarly effective. Combination of two serum assays can either enhance sensitivity or specificity, depending on the decision rules adopted, but no further discriminatory gain results from the combination of the three indicators.

Progress in Clinical Biochemistry
and Medicine, Vol. 8

1 Introduction

Suspect liver diseases and suspect acute pancreatitis provide contrasting examples in the weight of laboratory information in arriving at an efficient diagnosis. In both these, as all other diagnostic inquiries, the pretest probability of having a given condition at first presentation per se defines a likelihood of arriving at the correct conclusion. To improve the chance of attaining this target, the physician can next utilize three broad information sources: (a) attendant circumstances, most important signs, symptoms and epidemiological information, (b) imaging and related biophysical procedures, and (c) laboratory assays performed on samples from the patient. Each piece of diagnostic information so obtained can be qualified by its sensitivity and specificity (prior probabilities). However, ultimate knowledge of these characteristics alone does not permit successful interpretation of findings. Rather, efficient diagnostic strategies must consider aggregate data sets provided by different available sources (joint probability), and ultimately must answer the question what is the probability a condition is present, given a certain set of findings, as well as its corollary, what is the probability a condition is absent, given a certain set of findings (posterior probability)[1,2].

2 Purposes of Liver Diagnosis and Criteria for their Evaluation

There are three basic purposes of liver diagnosis. First, differentiation of hemolytic, parenchymatous, and obstructive jaundice attempts to resolve the question whether medical or surgical treatment is appropriate. Second, the evaluation of disease cause and therapeutic success attempts to resolve the twin questions whether incipient or residual liver cell damage is present, and whether the disease process is acute or

Table 1. Traditional use of laboratory indicators for the three purposes of liver diagnosis (see text)

Parameter	Purpose		
	1	2	3
Bilirubin, Total and Direct	×	×	
Urine Urobilinogen, Bilirubin	×	×	
Bile Acids	×		
Alkaline Phosphatase, 5′-Nucleotidase	×	×	
Gamma-Glutamyl Transferase		×	×
Transaminases	×	×	×
Immunoglobulins		×	×
Albumin			×
Prothrombin Time, Partial Thromboplastin Time			×
BSP Elimination			×

chronic. Third, the evaluation of the nature of the liver lesion attempts to resolve the twin questions whether a diffuse or a localized process is present, and what the functional state of the liver is in the presence of hepatomegaly.

The traditional applications of laboratory indicators to attain these purposes are shown in Table 1, but such established relationships deserve to be reassessed according to explicit criteria. In the case of the first purpose, valid criteria of success are selectivity in differential diagnosis and etiologic selectivity (e.g., posthepatitic or nutritional cirrhosis). For the evaluation of the disease course of the second purpose, sensitivity is a valid criterion, and for the evaluation of the nature of hepatic lesions or the third purpose, specificity with regard to morphologic changes. To assess the validity of these criteria, we have applied them to three common disease entities: viral hepatitis, obstructive jaundice and hepatic metastases.

2.1 Viral Hepatitis

About three fourths of patients with an acute onset of jaundice will have this disease (Table 2). A variety of attendant circumstances such as crowded living conditions [3], hemophilia, status post renal transplantation [4], or recent transfusion increase this pretest probability. Serum bilirubin, transaminases and alkaline phosphatase are usually evaluated in the light of clinical findings to establish the diagnosis. No single test confirms the diagnosis with certainty, but transaminases are very sensitive if any elevation above the norm is accepted as a positive, and very specific if a high value is observed [5]. Neither the total nor the direct bilirubin assists differential diagnosis, but these two tests are of value in anicteric presentation and as a baseline for comparisons in the future course of the disease [6].

The relative degrees of the transaminase and alkaline phosphatase elevations are considered useful for differential diagnosis. For instance, if the former is increased over sixfold and the latter less than 2.5 times, nine in ten patients will have hepatitis, while the converse pattern produces the same odds in favor of obstructive jaundice [5,7]. However, this approach only provides a limited detection rate as few jaundiced patients fit the diagnostic combinations of findings. Therefore, expanded assay batteries interpreted by multivariate analysis have been tested to enhance diagnostic classi-

Table 2. Sensitivities and specificities of laboratory indicators for acute viral hepatitis (modified after [6])

Parameter	Sensitivity	Specificity	Ref.
Acute Onset of Jaundice	.75	N.A.	
Aspartate Aminotransferase			
> 200 U/L	.75	.80	5)
> 600 U/L	.50	.99	
>1000 U/L	.30	1.00	
HBsAg	.80	.97	10, 11)
IgM Anti-HBc	.90	.98	
Anti-HAV	.99	.85	13)
IgM Anti-HAV	.99	.99	

fication. Still, laboratory assays alone typically could only arrive at the correct diagnosis of liver disease in about one of two patients [8]. Consequently, clinical findings must remain an important component of the diagnostic process.

Confirmation of acute viral hepatitis is greatly strengthened by viral markers. A positive IgM anti-HAV assay is diagnostic of hepatitis A. A positive HBsAg assay is virtually diagnostic of hepatitis B, but a negative test does not rule out the diagnosis [9–11]. However, the latter can be confirmed by a positive follow-up IgM anti-HBc assay. Still, liver biopsy probably remains the diagnostic gold standard for acute viral hepatitis [12].

2.2 *Obstructive Jaundice*

About four in ten adults presenting with jaundice as the first disease manifestation will have extrahepatic obstruction [15–17] (Table 3). Advancing age, the presence of abdominal pain, or a palpable gallbladder raise this pretest probability, while findings suggestive of parenchymal liver disease diminish it. Serum bilirubin, alkaline phosphatase and transaminase along with urinary urobilinogen are usually evaluated in the light of clinical findings to establish the diagnosis. Neither the total nor the direct bilirubin unequivocally assist differential diagnosis. The latter indicator lacks specificity for obstructive jaundice, since direct bilirubin fractions of 40 to 60% occur in parenchymal jaundice as well [18]. On the other hand, absence of urinary urobilinogen is both a highly sensitive and specific finding [15].

Alkaline phosphatase is a better test to rule out obstruction than to rule it in, particularly if the elevation is less than threefold as often occurs in partial or intermittent obstruction [15]. Similarly, there is marked overlap in transaminase findings among patients with hepatocellular and with obstructive disease. Therefore, only near normal or markedly elevated transaminase findings enhance discrimination. Still, combined with clinical findings the described biochemical data allow distinction of these two types of jaundice in up to nine of ten cases [19,20].

Table 3. Sensitivities and specificities of laboratory indicators for obstructive jaundice (modified after [14])

Parameter	Sensitivity	Specificity	Ref.
Adult Presenting with Jaundice	.40	N.A.	15–17)
Abdominal Pain	.70	.70	15)
Relief of Pain by Flexion	.20	.98	15)
Palpable Gallbladder	.30	1.00	15)
Direct Bilirubin >50%	.95	.30	15)
Urine Urobilinogen Negative	1.00	.90	15)
Alkaline Phosphatase >3×	.85	.65	15)
Ultrasound	.85	.80	19, 21, 22)
Computerized Tomography	.90	.90	19, 21, 22)
Transhepatic Cholangiography	.95	.99	21, 23)
Endoscopic Retrograde Cholangiopancreatography	.95	.99	23, 24)

Confirmation of obstruction and elucidation of its cause and site necessitates imaging procedures. With a low pretest probability, the lower sensitivities of ultrasound or computerized tomography suffice to rule-out obstruction [19,21,22]. Conversely, a high pretest probability demands the higher sensitivities offered by transhepatic cholangiography or endoscopic retrograde cholangiopancreatography [21,23,24].

2.3 Hepatic Metastases

When diagnosed, between one- and two-thirds of patients with colorectal neoplasms have hepatic metastases [26,27], while lung or breast cancer are associated with a much lesser frequency. Detection of the metastases demands both high sensitivity to spare false negatives the morbidity of unnecessary surgery, and high specificity to avoid denying false positives curative surgery. Neither hepatic enzymes such as alkaline phosphatase or gamma-glutamyl transferase, nor cancer markers such as carcinoembryonic antigen, combine the required discriminatory properties (Table 4) [28-30]. Therefore, diagnosis of hepatic metastases ultimately rests on imaging procedures [31-33]. On the other hand, recognition of hepatic metastases poses lesser problems in the minority of patients in whom jaundice, hepatomegaly or a painful liver suggest inoperable disease, or in patients with pancreatic neoplasms, where initial diagnostic imaging includes visualization of the liver.

Table 4. Sensitivities and specificities of laboratory indicators for hepatic metastases (modified after [25])

Parameter	Sensitivity	Specificity	Ref.
Alkaline Phosphatase	.50–.80	.60–.75	28, 29)
Gamma-Glutamyl Transferase	.75	.45	28)
Carcinoembryonic Antigen	.85	.60	28, 29, 30)
Liver Nuclide Scan			
— Any Abnormality	.90	.70	31)
— Focal Lesions	.80	.90	31, 32, 33)
Ultrasound	.80	.90	32, 33)
Computerized Tomography	.90	.90	32, 33)

2.4 Capabilities of Different Methods Used in Liver Diagnosis

The described examples provide a reasonable basis to evaluate the role of laboratory tests according to the three basic purposes of liver diagnosis presented at the outset.

Testing of bilirubin metabolism in liver disease provides the following:

1) good capability for the differential diagnosis of jaundice,
2) limited use for the evaluation of the severity and course of hepatitis (e.g., anicteric hepatitis), and
3) uninformative data concerning the morphology of the liver lesion.

Testing of the serum enzyme pattern in liver disease provides the following:

1) excellent capability for differential diagnosis (e.g., alkaline phosphatase and leucine aminopeptidase are elevated in obstruction, transaminases are elevated in hepatitis),
2) very good capability for the evaluation of the course of liver disease (e.g., early detection of hepatitis in blood donors, evaluation of completeness of recover) but of limited use for the evaluation of disease severity (e.g., sometimes no transaminase elevation in coma), and
3) some capability of providing information on the morphology of liver lesions (e.g., organelle specificity of enzymes).

The differential diagnostic capabilities of combining assays of different enzymes is probably best illustrated by the hepatotoxic effects of drugs. Figure 1 shows how the simultaneous assessment of just two enzyme indicators, a transaminase and alkaline phosphatase, can distinguish the multiple pathologies which characterize the toxicities of different drugs [34]. Clearly, not only the fact that the indicator enzymes are within normal limits or elevated, but the additional circumstance of their relative elevation when compared to each other provides diagnostic information.

In summary, we thus can conclude:

1) Differential diagnosis rests, on the one hand, on such clinical findings as history, palpation, and findings on other organs or systems, and, on the other hand, on liver function tests.

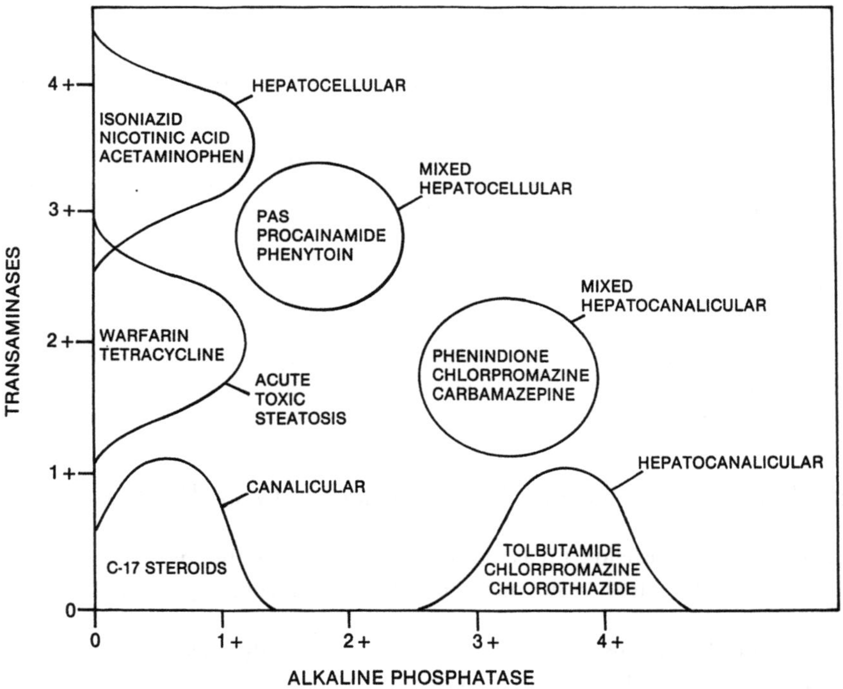

Fig. 1. Discrimination of hepatic side effects of drugs through the simultaneous assay of alkaline phosphatase and aspartate aminotransferase or alamine aminotransferase. Selected examples are given for each type of hepatic reaction (modified after [34]).

2) For the evaluation of the disease course, liver function tests are ideal.
3) For the evaluation of the nature of hepatic lesions, laparoscopy and liver biopsy are ideal, imaging procedures offer an alternative, and liver function tests at best provide supportive findings.

3 Diagnosis of Acute Pancreatitis

Chronic alcoholism and biliary tract disease are the twin main causes of acute pancreatitis. The former most commonly predisposes men in the third and fourth decade, the latter women in the fifth and later decades. Abdominal surgery, trauma, penetrating peptic ulcer, drugs, diabetes mellitus, anemia, systemic infections, hyperlipemia, hyperparathyroidism are other risk factors which increase the pretest probability of the condition. In almost half of all cases, recurrences occur at intervals. The relatively sensitive clinical findings include abdominal pain (90%), nausea and vomiting (85%), flatulence (80%), fever (80%), and hypotension (65%), but all these indicators lack specificity [35]. As a consequence, efficient diagnosis crucially relies on laboratory findings.

The test armamentarium of enzyme indicators for acute pancreatitis continues to improve. Not only have new modifications substantially enhanced the reliability of such standbys as amylase and lipase assay, but less traditional indicators such as trypsinogen and elastase assay find expanding clinical use. Clearly, the widening analytical choices pose the question of what testing scheme provides the most effective diagnostic strategy for acute pancreatitis. This in turn raises the issue of what constitutes a realistic assessment of the diagnostic performance by a given approach to testing.

3.1 Conceptual Considerations

Valid trials, obviously, must duplicate as closely as possible the real life circumstances in which the testing is to be used. In this respect, most test evaluations appear deficient in at least one of three ways. First, the assessment should not compare findings in classical cases with those in healthy subjects. Rather, diagnostic decision should be evaluated on a mixed sample representing the population in which the tests are used. Second, diagnostic discrimination may have been evaluated based on peak values retrospectively selected from serial measurements, rather than based on the initial findings confronting the physician forced to diagnose an acute condition. Third, results from more than one indicator may not have been combined to evaluate the odds of a correct diagnosis in the face of conflicting findings.

We have proposed that the clinical validation of diagnostic assays be structured in analogy to therapeutic clinical trials, which by common understanding are divided into three phases [36,37]. Each consecutive phase of assessment would define increasingly stringent requirements for the three essential aspects of trials, (a) selection of the patient sample, (b) conditions for observations, and (c) statistical descriptors of results

Table 5. A Model to evaluate the performance of diagnostic clinical trials.
Requirements at three levels of stringency are given for three crucial design parameters

Sampling	
Phase I:	Correct Clinical Classification
Phase II:	Representative
Phase III:	To Reflect all Circumstances
Observations	
Phase I:	No Circular Logic
Phase II:	Unselected
Phase III:	Double Blind
Descriptors	
Phase I:	Prior Probabilities (Sensitivity, Specificity)
Phase II:	Posterior Probabilities (Bayes' Theoreum)
Phase III:	Expected Value

(Table 5). An evaluation of studies reported in the scientific literature according to this analytical model should answer two distinct questions: How good is a given test, and how well is the performance claimed for a given test documented?

The proposed model's use can be illustrated by a review of the existing information on two tests: assay of serum amylase isoenzymes with the help of the wheat germ inhibitor, and assay of serum trypsinogen. Quite apart from what diagnostic capabilities might be claimed for these assays, their documentation is widely disparate. The former test barely is validated at the level of the first phase of our model in the four studies we identified [38–41]. Their aggregate report findings in little more than twenty acute pancreatitis cases on which an estimate of sensitivity can be based. Thus, the claims for 100% sensitivity must be dismissed as poorly documented if not doubtful. Selection and documentation of the control cases, on the other hand, is altogether inadequate to define a clinically valid estimate of specificity. In marked contrast, more than half a dozen trials, involving in their aggregate over 200 cases of acute pancreatitis as well as a much larger number of well chosen and defined controls, document the performance of the latter test [42–49]. Thus, the claims of 88–100% sensitivity and 89–100% specificity can be considered adequately validated even at phase II of the model.

3.2 A Clinical Trial

To assess the discriminatory capability of laboratory indicators of acute pancreatitis, we used assays of urinary amylase and of serum amylase, lipase, trypsinogen and elastase in the same group of 67 patients in whom acute pancreatitis was a diagnostic consideration [50]. Diagnostic classification established after the patients' discharge by review of all clinical findings, including history and laboratory data, but exclusive of the investigated enzyme indicators, produced a group of 33 index cases (18 male), and 34 controls (15 male) including patients with peptic ulcer, cholecystitis, small bowel obstruction, appendicitis, gastritis and various other conditions. The trial was designed to meet phase I and II requirements and possibly even phase III requirements with regard to both the criteria for sampling and observations. However, it is still necessary to caution that in estimating the correctness of test outcome, our purpose was not to establish absolute measures of test capability, but only the relative measures obtained with different strategies, in order to define the best diagnostic approach among several.

Table 6. Sensitivities and specificities of enzyme indicators of acute pancreatitis

Indicator	Initial Values		Highest Values	
	Sensitivity	Specificity	Sensitivity	Specificity
Urine Amylase — Activity per Volume	.59	.84	.70	.78
— Activity Excreted per Hour	.66	.83	.73	.72
— Amylase/Creatinine Clearance	.63	.55	.59	.68
Serum Amylase	.82	.94	.91	.94
Serum Elastase	.91	1.00	.94	.94
Serum Lipase	.91	1.00	.94	.97
Serum Trypsinogen	.91	.93	.91	.87

Table 6 lists sensitivities and specificities for individual tests in two different circumstances. Discrimination based on initial values obtained immediately upon a patient's admission is compared to discrimination based on highest individual values identified retrospectively among repeated measurements made during a patient's hospitalization. Urine amylase assays, analyzed in three frequently recommended ways (activity per volume, activity excreted per hour, amylase/creatinine clearance ratio), consistently produced relatively poor discrimination. All serum assays provided markedly better diagnostic classification, with the sensitivities of initial values exceeding 0.80 and their specificities 0.90. Using the highest values instead did not convincingly improve discrimination, as a small gain in sensitivity (reduction in false negatives) was simply traded for a similar loss in specificity (increase in false positives).

In an attempt to neutralize the effects of univariate false positives and negatives and thus enhance classification, it is possible to combine two or more discriminators. When the six pairs of tests possible with four indicators were analyzed demanding that both tests be positive in order to diagnose acute pancreatitis (Boolean intersection or the "and" strategy), specificity reached 1.00 with all pairs, whether initial or highest values were used. However, the price for this improvement was some loss in sensitivity. Conversely, when pairs of tests were analyzed accepting only one positive in order to diagnose acute pancreatitis (Boolean union on the "or" strategy), sensitivity increased reaching 1.00 for the highest values of some test pairs, but specificity decreased in all instances except for the initial values of one test pair. No obvious winner or loser combination of indicators emerged. Still, in all evaluated circumstances the best performing test pair included lipase and the worst performing amylase.

Four different indicator triplets could be formed. Just as expected, the highest findings showed a smaller sensitivity of all "and" triplets than that obtained with indicator pairs, while the "or" triplets reduced specificity. These reductions were not as clearcut for initial findings, and the discriminatory properties of triplets fell roughly into the domains also occupied by bivariate diagnosis. In all tested conditions, the combination lipase, elastase and trypsinogen provided the greatest sum of sensitivity and specificity, but in no instance did this triplet outperform the best pair of its component indicators.

4 References

1. Werner M, Brooks SH, Wetter R (1973) Hum Pathol 4: 17
2. Werner M (1978) in: Benson ES, Rubin M (eds) Logic and economics of clinical laboratory use. Elsevier, New York, p 41
3. Lemon SM, Lednar WM, Bancroft WH et al (1982) Am J Epidemiol 116: 438
4. Ware AJ, Luby JP, Hollinger B et al (1979) Ann Intern Med 91: 364
5. Clermont RJ, Chalmers TC (1967) Medicine 46: 197
6. Holt JT, Arvan DA (1986) in: Griner PF, Panzer RJ, Greenland P (eds) Clinical diagnosis and the laboratory. Year Book 1986, Chicago, p 270
7. Shearman DJC, Finlayson NDC (1982) in: Shearman DJC, Finlayson NDC (eds) Diseases of the gastrointestinal tract and liver. Churchill Livingston, New York, p 451
8. Sher P (1977) Clin Chem 23: 627
9. Hoofnogle JH (1983) Hepatol 3: 267
10. Chau KH, Hargie MP, Decker RH et al (1983) Hepatol 3: 142
11. Perrillo RP, Chau KH, Oveby LR et al (1983) Gastroenterol 85: 163
12. DeRitis F, Giusti G, Piccinino F et al (1973) Acta Hepato-Gastroenterol. 20: 371
13. Storch GA, Bodicki C, Parker M et al (1982) Am J Med 73: 663
14. Wilbur DC, Arvan DA (1986) in: Griner PF, Panzer RG, Greenland P (eds) Clinical diagnosis and the laboratory. Year Book 1986, Chicago, p 256
15. Schenker S, Balint J, Schiff L (1962) Am J Dig Dis 7: 449
16. Knill-Jones RP et al (1973) Brit Med J 1: 530
17. Malchow-Moller A et al (1981) Scand J Gastroent 16: 1
18. Fevery J, Claes J, Heirwegh K et al (1967) Clin Chim Acta 17: 73
19. Thomas MJ, Pellegrini CA, Way LW (1982) Am J Surg 144: 102
20. Siegel JH, Yatto RP (1982) Arch Intern Med 142: 1877
21. Sample WF et al (1978) Radiol 128: 719
22. Baron RL et al (1982) Radiol 145: 91
23. Gold RP, Casarella WJ, Stern G et al (1979) Radiol 133: 39
24. Shimizu H et al (1981) Radiol 138: 411
25. Panzer RJ (1986) in: Griner PF, Panzer RG, Greenland P (eds) Clinical diagnosis and the laboratory. Year Book 1986, Chicago, p 284
26. Foster JH, Lundy J (1981) Curr Probl Surg 18: 157
27. Finlay IG et al (1982) Brit Med J 2: 803
28. Kemeny MM et al (1982) Ann Surg 195: 163
29. Tartter PI et al (1981) Ann Surg 193: 357
30. Szymendera JJ (1982) Dis Colon Rectum 25: 191
31. Christensen M (1982) Acta Med Scand 211: 23
32. Smith TJ, Kemeny MM, Sugarbaker PH et al (1982) Ann Surg 195: 486
33. Alderson PO, Adams DF, McNeil BJ et al (1983) Radiol 149: 225
34. Zimmerman HJ (1978) Hepatotoxicity. The adverse effects of drugs and other chemicals on the liver. Appleton-Century-Crofts, New York
35. Maeder F (1972) Fortschr Med 90: 572
36. Werner M (1985) Bull Mol Biol Med 10: 405

37. Werner M (1984) in: Burlina A, Galzigna L (eds) Clinical enzymology symposia 4. Piccin, Padua, Italy, p 313
38. O'Donnell MD, Fitz Gerald O, McGeeney KF (1977) Clin Chem 23: 560
39. Huang WY, Tietz NW (1982) Clin Chem 28: 1525
40. Tietz NW, Shuey DF (1984) Clin Chem 30: 1227
41. Okabe H, Uji Y, Netsu K, Noma A (1984) Clin Chem 30: 1219
42. Elias E, Redshaw M, Wood T (1977) Lancet 2: 66
43. Lake-Bakaar G, McKavanagh S, Gatus B, Summerfield JA (1980) Scand J Gastroent 15: 97
44. Fahrenkrug J, Magid E (1980) Clin Chem 26: 1573
45. Masoero G et al (1982) Digest Dis Sci 27: 1089
46. Ventrucci M et al (1983) Digestion 28: 114
47. Rinderknecht H, Adham NF, Renner IG, Abramson SB (1983) in: Goldberg DM, Werner M (eds) Progress in clinical enzymology, vol 2. Masson, New York, p 77
48. Steinberg WM et al (1985) Ann Intern Med 102: 576
49. Eckfeldt JH et al (1985) Arch Pathol Lab Med 109: 316
50. Werner M, Steinberg WM, Pauley C (1989) Clin Chem in press

Author Index Volumes 1–8

The volume numbers are printed in italics

Subject Index

Progress in Clinical Biochemistry and Medicine

Eds.: E. Baulieu, D. T. Forman, L. Jaenicke, J. A. Kellen, Y. Nagai, G. F. Springer, L. Träger, L. Will-Shahab, J. L. Wittliff

Volume 5

Regulation of Plasma Low Density Lipoprotein Levels · Biopharmacological Regulation of Protein Phosphorylation · Calcium-Activated Neutral Protease · Microbial Iron Transport · Pharmacokinetic Drug Interactions

1987. 59 figures. VII, 134 pages. ISBN 3-540-18187-3

Volume 4

Control of Immune Response by Endocrine Factors · Malaria Vaccine · Controlled Drug Delivery · Enzyme-Immunoassay

1987. 45 figures. 146 pages. ISBN 3-540-16955-5

Volume 3

Metabolic Control in Diabetes Mellitus · Beta Adrenoceptor Blocking Drugs · NMR Analysis of Cancer Cells · Immunoassay in the Clinical Laboratory · Cyclosporine

1986. 68 figures. VIII, 192 pages. ISBN 3-540-16249-6

Volume 2

Oncogenes and Human Cancer · Blood Groups in Cancer · Copper and Inflammation · Human Insulin

1985. 25 figures. VII, 163 pages. ISBN 3-540-15567-8

Volume 1

Essential and Non-Essential Metals · Metabolites with Antibiotic Activity · Pharmacology of Benzodiazepines · Interferon Gamma Research

1984. 42 figures. VII, 203 pages.
ISBN 3-540-13605-3

Springer-Verlag
Berlin Heidelberg New York
London Paris Tokyo Hong Kong

Springer

MIX
Papier aus verantwortungsvollen Quellen
Paper from responsible sources
FSC® C105338

If you have any concerns about our products,
you can contact us on
ProductSafety@springernature.com

In case Publisher is established outside the EU,
the EU authorized representative is:
Springer Nature Customer Service Center GmbH
Europaplatz 3, 69115 Heidelberg, Germany

Printed by Libri Plureos GmbH
in Hamburg, Germany